FUNDAMENTALS OF
Grid Computing

Theory, Algorithms and Technologies

CHAPMAN & HALL/CRC
Numerical Analysis and Scientific Computing

Aims and scope:
Scientific computing and numerical analysis provide invaluable tools for the sciences and engineering. This series aims to capture new developments and summarize state-of-the-art methods over the whole spectrum of these fields. It will include a broad range of textbooks, monographs, and handbooks. Volumes in theory, including discretisation techniques, numerical algorithms, multiscale techniques, parallel and distributed algorithms, as well as applications of these methods in multi-disciplinary fields, are welcome. The inclusion of concrete real-world examples is highly encouraged. This series is meant to appeal to students and researchers in mathematics, engineering, and computational science.

Proposals for the series should be submitted to one of the series editors above or directly to:
CRC Press, Taylor & Francis Group
4th, Floor, Albert House
1-4 Singer Street
London EC2A 4BQ
UK

Published Titles

Classical and Modern Numerical Analysis: Theory, Methods and Practice
Azmy S. Ackleh, Edward James Allen, Ralph Baker Kearfott,
* and Padmanabhan Seshaiyer*

A Concise Introduction to Image Processing using C++
Meiqing Wang and Choi-Hong Lai

Decomposition Methods for Differential Equations:
** Theory and Applications**
Juergen Geiser

Grid Resource Management: Toward Virtual and Services Compliant Grid
Computing
Frédéric Magoulès, Thi-Mai-Huong Nguyen, and Lei Yu

Fundamentals of Grid Computing: Theory, Algorithms and Technologies
Frédéric Magoulès

Introduction to Grid Computing
Frédéric Magoulès, Jie Pan, Kiat-An Tan, and Abhinit Kumar

Mathematical Objects in C++: Computational Tools in a Unified Object-
Oriented Approach
Yair Shapira

Numerical Linear Approximation in C
Nabih N. Abdelmalek and William A. Malek

Numerical Techniques for Direct and Large-Eddy Simulations
Xi Jiang and Choi-Hong Lai

Parallel Algorithms
Henri Casanova, Arnaud Legrand, and Yves Robert

Parallel Iterative Algorithms: From Sequential to Grid Computing
Jacques M. Bahi, Sylvain Contassot-Vivier, and Raphael Couturier

FUNDAMENTALS OF
Grid Computing
Theory, Algorithms and Technologies

Edited by
Frédéric Magoulès

CRC Press
Taylor & Francis Group
Boca Raton London New York

CRC Press is an imprint of the
Taylor & Francis Group, an **informa** business
A CHAPMAN & HALL BOOK

CRC Press
Taylor & Francis Group
6000 Broken Sound Parkway NW, Suite 300
Boca Raton, FL 33487-2742

First issued in paperback 2019

© 2010 by Taylor & Francis Group, LLC
CRC Press is an imprint of Taylor & Francis Group, an Informa business

No claim to original U.S. Government works

ISBN-13: 978-1-4398-0397-7 (hbk)
ISBN-13: 978-0-367-38460-9 (pbk)

Library of Congress Cataloging-in-Publication Data

Magoulès, F. (Frédéric)
    Fundamentals of grid computing : theory, algorithms and technologies / Frédéric Magoulès.
        p. cm. -- (Chapman & Hall/CRC numerical analysis and scientific computing)
    Includes bibliographical references and index.
    ISBN 978-1-4398-0367-7 (hardcover : alk. paper)
    1. Computational grids (Computer systems) I. Title. II. Series.

QA76.9.C58M339 2009
004'.36--dc22   2009038022

Visit the Taylor & Francis Web site at
http://www.taylorandfrancis.com

and the CRC Press Web site at
http://www.crcpress.com

Contents

List of Figures

List of Tables

Foreword

It is really a pleasure for me to write the foreword for this book, *Fundamentals of Grid Computing: Theory, Algorithms and Technologies.*

Grid computing is now becoming a very powerful and innovative tool allowing tens of thousands of researchers around the world to perform breakthroughs in their research projects. The most striking example can be seen from the most ambitious research project in the world, the Large Hadron Collider (LHC) giant accelerator at the European center for particle physics (CERN) in Geneva, which is relying completely on grid technology to store, process and analyze its huge volumes of distributed data. Just ten years ago, the LHC computing model was still based on shipping cassettes from one center to another through trucks and planes! This shows the fantastic progress made by grid technology during that time scale, and the maturity reached by these techniques is quite well-reflected in Frédéric Magoulès's book, which describes in a very clear way the state of the art grid middleware.

This very fast evolution became possible because of a fortunate quadruple coincidence in the early 2000s: the dramatic increase of high speed network links at affordable costs, the existence of a strong scientific community having a desperate need for solving what then seemed an insurmountable challenge in terms of computing g-cycles and data storage, the widespread diffusion of cheap Linux-based clusters and the availability of the grid toolkit, Globus. This fortunate situation led to the development of very ambitious grid production projects both in Europe and in the United States.

The European project enabling grids for e-science (EGEE) has now become a real production infrastructure running 24 hours a day and several million jobs per month thanks to its 250 nodes totaling 100,000 processors and 50 petabytes of storage. More importantly, more than a dozen scientific disciplines are using it for their advanced research work, ranging from particle physics, astronomy, life science, earth science, human sciences, medicine, chemistry, and finance to even art. A recent survey performed at the French national scale among more than 3,000 researchers from all scientific fields showed that the grids are, for the years to come, an essential tool in great complement to supercomputers.

This book shows, in some sense, the way to the future, where next generation middleware such as those described here will replace in the production infrastructure the more rudimentary ones in use today. Therefore, I am sure that the readers will greatly benefit from this insightful journey in the heart of the grids, a key technology in a very large number of scientific endeavors.

Guy Wormser
CNRS Institut des Grilles, France

Preface

The term "the grid" has emerged in the mid 1990s to denote a proposed distributed computing infrastructure which focuses on large-scale resource sharing, innovative applications, and high performance orientation. The grid concept is motivated by a real and specific problem: the coordinated resource sharing and problem solving of dynamic, multi-institutional, virtual organizations. The sharing is not primarily a file exchange but rather a direct access to computing resources, software, storage devices and other resources, with a necessary, highly controlled sharing rule which defines clearly and carefully what is shared, who is allowed to share and the conditions under which sharing occurs. Over the years, a combination of technology trends and research progress has resulted in an increased focus on grid technology for industry, commerce and business areas. Nowadays, grid technology has evolved toward open grid services architecture, in which a grid provides an extensible set of services.

This edited book follows the two previous authored books on grid computing published by Chapman & Hall/CRC Press in this series entitled: *Introduction to Grid Computing* by Frédéric Magoulès, Jie Pan, Kiat-An Tan, Abhinit Kumar (2009), and *Grid Resource Management* by Frédéric Magoulès, Thi-Mai-Huong Nguyen, Lei Yu (2008).

The main topics considered in the present book include: sharing resources, data replication, data management, fault tolerance, scheduling, broadcasting and load balancing algorithms. The nine chapters of this book are followed by two appendices introducing two types of software written in Java programming language. The first software deals with the implementation of some replications strategies for data replication in the grid. The second software deals with the implementation of a simulator for distributed scheduling in grid environments. These "easy-to-learn, easy-to-use" open source software allow the reader to get familiar with the grid technology covered in the previous chapters.

The various technology presented in this book demonstrates the wide aspects of interest in grid computing, and the many possibilities and venues that exist in the research in this area. We are sure that this interest is only going to further evolve, and that many exciting developments are still awaiting us.

Frédéric Magoulès
Ecole Centrale Paris, France

Preface

Warranty

Every effort has been made to make this book as complete and as accurate as possible, but no warranty of fitness is implied. The information is provided on an as-is basis. The authors, editor and publisher shall have neither liability nor responsibility to any person or entity with respect to any loss or damages arising from the information contained in this book or from the use of the code published in it.

Chapter 1

Grid computing overview

Frédéric Magoulès

Applied Mathematics and Systems Laboratory, Ecole Centrale Paris, Grande Voie des Vignes, 92295 Châtenay-Malabry, France

Thi-Mai-Huong Nguyen

Applied Mathematics and Systems Laboratory, Ecole Centrale Paris, Grande Voie des Vignes, 92295 Châtenay-Malabry, France

Lei Yu

Applied Mathematics and Systems Laboratory, Ecole Centrale Paris, Grande Voie des Vignes, 92295 Châtenay-Malabry, France

1.1 Introduction

The term "the grid" has emerged in the mid 1990s to denote a proposed distributed computing infrastructure which focuses on large-scale resource sharing, innovative applications, and high-performance orientation [Foster et al., 2001]. The grid concept is motivated by a real and specific problem – the coordinated resource sharing and problem solving of dynamic, multi-institutional virtual organizations. The sharing is not primarily a file exchange but rather a direct access to computing resources, software, storage devices, and other resources, with a necessary, highly controlled sharing rule which defines clearly and carefully just what is shared, who is allowed to share, and the conditions under which sharing occurs. A set of individuals and/or institutions defined by such sharing rules forms what we call a virtual organization (VO).

Now, a combination of technology trends and research progress results in an increased focus on grid technology in industry and commercial domain.

Grid technology is evolving toward an open grid services architecture (OGSA) in which a grid provides an extensible set of services that virtual organizations can aggregate in various ways. Building on concepts and technologies from both the grid and web services communities, OGSA defines a series of standards and specifications which supports the creation of grid service with location transparency and underlying native platform facilities [Foster et al., 2002b].

1.2   Definitions

Grids have moved from the obscurely academic to the highly popular. The growing need of the grid in commercial and scientific domain requests a clear definition of the word grid. The earliest definition of a grid emerged in 1969 by Len Kleinrock:

> "We will probably see the spread of 'computer utilities,' which, like present electric and telephone utilities, will service individual homes and offices across the country."

Ian Foster suggests a grid checklist in his paper [Foster, 2002] to identify a real grid system. The suggestion can be concluded into three points:

- A grid integrates and coordinates resources and users that live within different control domains. Current internet technologies address communication and information exchange among computers but do not provide integrated approaches to the coordinated use of resources at multiple sites for computation. Moreover, current distributed computing technologies such as CORBA, DCE, and Enterprise Java do not accommodate the range of resource types or do not provide the flexibility and control on sharing relationships needed to establish VOs. Fronting the problems above, grid technologies integrate different administrative units of the same company or different companies and address the issues of security, policy, payment, membership, and so forth.

- A grid is built from multi-purpose protocols and interfaces that address such fundamental issues as authentication, authorization, resource discovery, and resource access. In a large-scale grid environment, each resource is integrated from multiple institutions, each with their own policies and mechanisms. Thus it is important that these protocols and interfaces should be standard and open. Otherwise, we are dealing with an application specific system.

- A grid allows its constituent resources to be used in a coordinated fashion to deliver various qualities of service, relating for example to response time, throughput, availability, and security, and/or co-allocation of multiple resource types to meet complex user demands, so that the utility of the combined system is significantly greater than that of the sum of its parts.

Here the difference between two concepts "a grid" and "the grid" is also important to be distinguished. The grid vision requires protocols (interfaces and policies) that are not only open and general-purpose but also standard. This is these standards that allow us to establish resource-sharing arrangements dynamically with any interested party and thus to create a compatible and interoperable distributed systems.

According to the discussion above, we mix the definition in the book [Foster and Kesselman, 1998] with the VO concept, and denote:

> "A computational grid is a hardware and software infrastructure that provides dependable, consistent, pervasive, and inexpensive access to coordinated and shared resources in dynamic, multi-institutional virtual organizations. The sharing is not primarily file exchange but rather direct access to computers, software, data, and other resources. The sharing rule is clearly and carefully defined, enabling a necessary and high control of resources."

1.3   Classifying grid systems

Typically, grid computing systems are classified into computational and data grids. In the computational grid, the focus lies on optimizing execution time of applications that requires a great number of computing processing cycles. On the other hand, the data grid provides the solutions for large scale data management problems. In [Krauter et al., 2002], a similar taxonomy for grid systems is presented, which proposes a third category, the service grid.

Computational grid refers to systems that harness machines of an administrative domain in a "cycle-stealing" mode to have higher computational capacity than the capacity of any constituent machine in the system.

Data grid denotes systems that provide a hardware and software infrastructure for synthesizing new information from data repositories that are distributed in a wide area network.

Service grid refers to systems that provide services that are not provided by any single local machine. This category is further divided as on demand (aggregate resources to provide new services), collaborative (con-

nect users and applications via a virtual workspace), and multimedia
(infrastructure for real-time multimedia applications).

1.4   Grid applications

A grid is considered to be an infrastructure that bonds and unifies globally
remote and diverse resources in order to provide computing support for a
wide range of applications. The different types of computing offered by grids
can be categorized according to the main challenges that they present from
the grid architecture point of view. The types of computing are concluded as
follows [Bote-Lorenzo et al., 2003 , Foster and Kesselman, 1999]:

Distributed supercomputing: This type of computing allows applications to use grids to aggregate computational resources in order to reduce the completion time of a job or to tackle problems that cannot be
solved on a single system. The technical challenges include the need to
coschedule scarce and expensive resources, the scalability of protocols
and algorithms to tens or hundreds of thousands of nodes, the design
for latency-tolerant algorithms, achieving and maintaining high performance computing across heterogeneous systems.

High-throughput computing: In high-throughput computing, the grid
is used to schedule large numbers of loosely coupled or independent
tasks, with the goal of putting unused processor cycles often from idle
workstations to work. Chip design and parameter studies are normally
applications of this type of computing.

On-demand computing: On-demand applications use grid capabilities to
couple remote resources into local applications in order to fulfill short-
term requirements. These resources cannot be cost-effective or conve-
niently located and it may be computation, software, data repositories,
specialized sensors, and so on. The challenging issues in on-demand
applications derive primarily from the dynamic nature of resource requirements and the potentially large populations of users and resources.
These issues include resource location, scheduling, code management,
configuration, fault tolerance, security, and payment mechanisms.

Data-intensive computing: Data-intensive applications analyze and treat
information and data which are maintained in geographically distributed
repositories, digital libraries and databases, and aggregated by grid capabilities. Modern meteorological forecasting systems which make extensive use of data assimilation to incorporate remote satellite observations and high-energy physics are typical applications of data-intensive

computing. The challenge in data-intensive applications is the scheduling and configuration of complex, high-volume data flows through multiple levels of hierarchy.

Collaborative computing: In collaborative computing, applications are concerned primarily with enabling and enhancing human-to-human interactions. Such applications often provide a virtual shared space and are concerned with enabling the shared use of computational resources such as data archives and simulations. Challenging issues of collaborative applications from a grid architecture perspective are the realtime requirements imposed by human perceptual capabilities and the rich variety of interactions that can take place.

1.5   Grid architecture

A grid has a layered, extensible, and open architecture (Figure 1.1) which facilitates the identification for general classes of components. Components within each layer share common characteristics but can be built on capabilities and behaviors provided by any lower layer [Foster et al., 2001].

Fabric layer: The grid fabric layer provides interfaces to local control of resources which may be logical entities, computer clusters, or distributed computer pools. Fabric components make resources virtual by implementing the local, resource-specific operations that occur on specific resources (whether physical or logical). There is thus a interdependence between the functions of resource implemented at the fabric level and the shared operations supported. The principal resources which the fabric layer supports and operations of these resources are shown as follows: (1) *Computational resources:* Operations are required for starting programs and for monitoring and controlling the execution of the resulting processes. Management mechanisms are needed to control the resources and inquiry functions are required for determining hardware and software characteristics as well as relevant state information such as current load and queue state in the case of scheduler-managed resources. (2) *Storage resources:* Putting and getting files operations and management mechanisms are required. Inquiry functions are needed for determining hardware and software characteristics as well as relevant load information such as available space and bandwidth utilization. (3) *Network resources:* Management mechanisms and inquiry functions should be provided to control network transfers and to determine network characteristics and load. (4) *Code repositories:* Management mechanisms for managing versioned source and object code are needed. (5) *Catalogs:* Catalog query and update operations must be implemented, for example: a relational database.

Connectivity layer: The core communication and authentication protocols required for grid-specific network transactions are defined at the connectivity layer which enables the exchange of data between fabric layer resources. The identity of users and resources is verified by authentication protocols which have the following characteristics: single sign on, delegation, integration with various local security solutions, and user-based trust relationships.

Resource layer: The resource layer defines protocols (APIs and SDKs) for the secure negotiation, initiation, monitoring, control, accounting, and payment of sharing operations on individual resources. There are two primary classes of resource layer protocols: Information protocols and management protocols. Information protocol are used to obtain information about the structure and state of a resource and management protocols are used to negotiate access to a shared resource. Resource layer protocols are concerned entirely with individual resources and issues of global state and atomic actions across distributed collections will be discussed in the collective layer.

Collective layer: The collective layer contains protocols and services (APIs and SDKs) that are not associated with any specific resource but rather are global in nature and capture interactions across collections of resources. A wide variety of sharing behaviors and operations is implemented, such as directory services, monitoring and diagnostics services, data replication services, etc.

Applications layer: The applications layer comprises the user applications that operate within a VO environment.

1.6   Grid computing projects

There are many international grid projects, which are classified into several groups according to their positioning and functionality in a grid community.

1.6.1   Grid middleware (core services)

1.6.1.1   Globus

The Globus project which provides a open source Globus Toolkit is a fundamental enabling technology for the "grid," letting people share computing

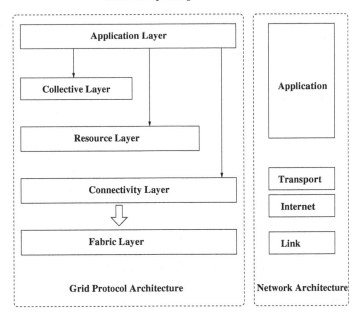

FIGURE 1.1: The layered grid architecture and the Internet protocol architecture.

power, databases, and other resources across multiple independent administrative domains without sacrificing local autonomy. The toolkit includes software services and libraries for resource monitoring, discovery, and management, plus security and file management.

The Globus Toolkit (GT) has been developed since the late 1990s and provides basic services and capabilities that are required to construct a computational grid. The toolkit consists of a set of modules. Each module defines an interface which can be invoked by higher-level services and provides an implementation which uses appropriate low-level operations to achieve user requested services [Foster and Kesselman, 1997].

The recent web services-based GT4 release provides significant improvements over previous releases in terms of robustness, performance, usability, documentation, standards compliance, and functionality [Foster, 2006]. The components of GT4 can be categorized into three groups:

- A set of services implementations: Most of these services are Java web services and they address execution management, data access and movement, replica management, monitoring discovery, credential management, and instrument management.

- Three containers: These containers can be used to host user-developed services written in Java, Python, and C. As services hosting environments, these containers provide implementations of security, management, discovery, state management and other mechanisms which are required by building services.

- A set of client libraries: These libraries allow client programs in Java, C, and Python to invoke operations in both GT4 and user-developed services.

1.6.1.2 gLite

gLite is a grid middleware that has been developed by the EGEE (Enabling Grids for E-SciencE) project, and is widely deployed among almost 50 European and other countries. The gLite grid services follow a service oriented architecture which facilitates the interactions between applications encapsulated by grid services, and which is compliant with upcoming grid standards, for instance the web service resource framework (WSRF) from OASIS and the open grid service architecture (OGSA) from the global grid forum. The gLite has a modular architecture, allowing users to deploy different services according to their needs, rather than being forced to use the whole system. This is intended to allow each user to tailor the system to their individual situation [Nakada et al., 2007].

gLite system consists of several modules which are presented as follows:

- Workload management system (WMS): WMS is a scheduling module for gLite. It receives job submission requests from users and allocates resources to jobs based on the resource information. WMS uses ClassAd [Raman et al., 1998] to make matchmaking and it can adopt different scheduling policies. For a submission request, if resources that match the job requirements are not immediately available, the request is kept in the task queue (TQ). Nonmatching requests will be retried either periodically or as soon as notifications of available resources appear.

- Berkeley directory information index (BDII): BDII is the information service module for gLite, which provides information access interface based on light-weight directory access protocol (LDAP).

- Computing element (CE): This is the job manager module for gLite. CE supports its own job description language (JDL) format, different from the JDSL standard.

1.6.1.3 Legion

Legion, an object-based metasystems software project at the University of Virginia, is designed for a system of millions of hosts and trillions of objects tied together with high-speed links. As a grid infrastructure, Legion presents users a view of a grid as a single virtual machine. This view reduces the complexities a user encounters before running applications or collaborating on a grid. The design principles of object-basedness and integration have enabled Legion to be extended and configured in a number of ways, while ensuring that the cognitive burden on the grid community is small. Groups of users can construct shared virtual work spaces, collaborate their researches, and exchange information in a grid community [Natrajan et al., 2001].

Legion has many features which traditional operating systems have provided, such as a single namespace, a file system, security, process creation and management, interprocess communication, input-output, resource management and accounting. In addition, features which are required in a grid system are also provided by Legion, such as complexity management, wide-area access, heterogeneity management, multi-language support, and legacy application support. The main features of Legion are described as follows:

- Object-basedness: Object-based design offers three advantages. First, it leads to a modular design which facilitates the complexity management of components. Second, it enables extending functionality by designing specialized versions of basic objects. Third, it enables selecting an intuitive boundary around an object for enforcing security. Although object-basedness is an essential feature in the design of Legion, grid users do not have to conform to object-based or object-oriented design. Legacy applications can be easily integrated by using the provided C++, C, Java, and Fortran interfaces without requiring any change to source or object code.

- Naming transparency: Every Legion object is assigned an identity (LOID). The LOID of an object is a sequence of bits that identifies the object uniquely in a given grid (and also across different grids) and it can be used to query all the information about this object, such as about its physical location, its current status, the permissions on it, associated metadata and the kind of service it provides (its interface). Once an object's interface is known, it can be requested to perform a desired service.

- Security: The authentication and access control lists (ACLs) for authorization in Legion is based on a public key infrastructure (PKI). Legion requires no central certificate authority to determine the public key of a named object because the object's LOID contains its public key.

- Integration: Legion provides a global, distributed file system for every grid it manages. This file system can contain any Legion object, such as other file systems, files, machines, users, console objects, and applications. Therefore this distributed file system enables richer collaboration than the internet or the web. Moreover, Legion provides a suite of high performance computing tools for running legacy applications, MPI applications, and PVM applications.

1.6.1.4 UNICORE

In 1997, the development of the uniform interface to computing resources (UNICORE) system was initiated to enable German supercomputer centers to provide their users with a seamless, secure, and intuitive access to their heterogeneous computing resources [Streit et al., 2005]. At the beginning UNICORE was developed as a prototype software in two projects funded by the German research ministry (BMBF). Over the following years, UNICORE evolved to a full-grown and well-tested grid middleware system, which today is used in daily production at many supercomputing centers worldwide.

UNICORE meets the open grid services architecture (OGSA) concept and all its components are implemented in Java. UNICORE has a layered grid architecture which consists of user, server, and target system tier.

- User tier provides a graphical user interface to exploit the entire set of services offered by the underlying servers. The client communicates with the server tier by sending and receiving abstract job objects (AJO) and file data via the UNICORE protocol layer (UPL) which is placed on top of the SSL protocol. AJOs are sent to the UNICORE gateway in form of serialized and signed Java objects and it contains platform and site independent descriptions of computational and data related tasks, resource information, and workflow specifications along with user and security information.

- Server tier controls the access to a UNICORE site and provides the virtualization of the underlying resources by mapping the abstract job on a specific target system. Each participating organization (e.g., a supercomputing center) to the grid is identified into a Usite with a symbolic name. A Usite consists of Vsites which represent resources in a Usite and support resources with different system architectures (e.g., a single supercomputer or a Linux cluster with resource management system).

- Target system tier implements the interface to the underlying computing resource with its resource management system. It is a stateless daemon running on the target system and interfacing with the local resource manager (e.g., PBS [PBS, 2006] or GRAM [Foster, 2006]).

During the development of the UNICORE technology, lots of European and international projects have decided to base their grid software implementations on UNICORE or to extend the growing set of core UNICORE functions in their projects as new features specific. Now the UNICORE software is available as open source which encourages the growing of developers community of core UNICORE and makes future development efforts open to the public.

1.6.2 Grid resource brokers and schedulers

During the second generation, we saw the tremendous growth of grid resource brokers and scheduler systems. The primary objective of these systems is to couple commodity machines in order to achieve the equivalent power of supercomputers with a significantly less expensive cost. A wide variety of powerful grid resource brokers and scheduler systems, such as Condor, PBS, Maui scheduler, LSF, and SGE spread throughout academia and business.

1.6.2.1 Condor

The Condor project [Condor, 2009] developed at the University of Wisconsin-Madison introduces the Condor high throughput computing system, which is often referred to simply as Condor and Condor-G.

- The Condor high throughput computing system [Tannenbaum et al., 2001] is a specialized workload management system for executing computer intensive jobs on a variety of platform environments (i.e., Unix and Windows). Condor provides a job management mechanism, scheduling policy, priority scheme, resource monitoring, and resource management. The key feature of Condor is the ability to scavenge and manage wasted CPU power from idle desktop workstations across an entire organization. Workstations are dynamically placed in a resource pool whenever they become idle and removed from the resource pool when they get busy. Condor is responsible for allocating a machine from the resource pool for the execution of jobs and monitoring the activity on all the participating computing resources.

- Condor-G [Frey et al., 2002] is the technological combination of the Globus and the Condor projects, which aims to enable the utilization of large collections of resources spanning across multiple domains. Globus contribution composes of the use of protocols for secure inter-domain communications and standardized access to a variety of remote batch systems. Condor contributes with the user concerns of job submission, job allocation, error recovery, and creation of a user-friendly environment. Condor technology provides solutions for both the frontend and backend of a middleware as shown in the Figure 1.2. Condor-G offers an interface for job reliable submission and management for the whole system. The Condor high throughput computing system can be used as the fabric management service for one or more sites. The Globus toolkit can be used as the bridge interfacing between them.

1.6.2.2 Portable batch system

The portable batch system (PBS) project [PBS, 2006] is a flexible batch queuing and workload management system originally developed by Veridian

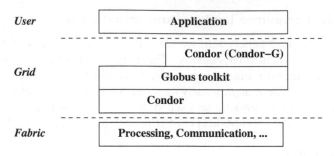

FIGURE 1.2: Condor in conjunction with Globus technologies in grid middleware, which lies between the user's environment and the actual fabric (resources) [Thain et al., 2005].

Systems for NASA. The primary purpose of PBS is to provide controls for initiating and scheduling the execution of batch jobs. PBS operates on a variety of networked, multi-platform UNIX environments, from heterogeneous clusters of workstations to massively parallel systems. PBS supports both interactive and batch mode, and provides a friendly graphical interface for job submission, tracking, and administrative purposes.

PBS is designed based on client-server model. The main components are *pbs_server* server process, which manages high-level batch object such as queues and jobs and *pbs_mom* server process, which is responsible for job execution. The *pbs_server* receives submitted jobs from users in the form of a script and schedules the job for later execution by a *pbs_mom* process.

PBS consists of several built-in schedulers, each of which can be customized for specific requirements. The default scheduler in PBS maximizes the CPU utilization by applying the first in first out (FIFO) method. It loops through the queued job list and starts any job that fits in the available resources. However, this effectively prevents large jobs from ever starting since the required resources are unlikely to ever be available. To allow large jobs to start, this scheduler implements a "starving jobs" mechanism, which defines circumstances under which starving jobs can be launched (e.g., first in the job queue, waiting time is longer than some predefined time). However, this method may not work under certain circumstances (e.g., the scheduler would halt starting of new jobs until starving jobs can be started). In this context, Maui scheduler has been adopted as plug-in scheduler for PBS system.

1.6.2.3   Maui scheduler

The Maui scheduler [Bode et al., 2000] developed principally by David Jackson for the Maui High Performance Computer Center is an advanced batch job scheduler with a large feature set, well suited for high performance computing (HPC) platforms. The key to the Maui scheduling design is its wall-time based reservation system, which allows sites to control exactly when, how, and by whom resources are used. The jobs are queued and managed based upon its priority, which is specified from several configurable parameters.

Maui uses a two-phase scheduling algorithm. During the first phase, the scheduler starts jobs with highest priority and then makes a reservation in the future for the next high priority job. In the second phase, Maui scheduler uses the backfill mechanism to ensure that large jobs (i.e., starving jobs) will be executed at a certain moment. It attempts to find lower priority jobs that will fit into time gaps in the reservation system. This gives large jobs a guaranteed start time, while providing a quick turn around for small jobs. In this way, the resource utilization is optimized and job response time is minimized. Maui uses the fair-share technique when making scheduling decisions based on job history.

1.6.2.4 Load sharing facility

Load sharing facility (LSF) [Platform, 2009] is a commercial resource manager for cluster from Platform Computing Corporation. It is currently the most widely used commercial job management system. LSF design focuses on the management of a broad range of job types such as batch, parallel, distributed, and interactive. The key features of LSF include system supports for automatic and manual checkpoints, migrations, automatic job dependencies, and job re-schedulings.

LSF supports numerous scheduling algorithms, such as first come first served, fair-share, backfill. It can also interface with external schedulers (e.g., Maui), which complement features of the resource manager and enable sophisticated scheduling.

1.6.2.5 Sun grid engine

Sun grid engine (SGE) [Sun, 2009b] is a popular job management system supported by Sun Microsystems. It supports distributed resource management and software/hardware optimization in heterogeneous networked environments.

A user submits a job to the SGE, together with the requirement profile, user identification, and a priority number for the job. The requirement profile contains attributes associated with the job, such as memory requirements, operating system required, available software licenses, etc. Then, jobs are kept waiting in a holding area until resources become available for execution. Based on the requirement profile, SGE assigns the job to an appropriate queue associated with a server node on which the job will be executed. SGE maintains load balancing by starting new jobs on the least loaded queue to spread workload among available servers.

1.6.3   Grid systems

1.6.3.1   GridLab

In 2000, the Applications Research Group (APPS-RG) of the GGF established a pan-European testbed, based on the Globus Toolkit, for prototyping and experimenting with various application scenarios. Based on these testbed experiences, the European Commission launched an application oriented project, called GridLab [Allen et al., 2003]. The primary goal of GridLab is to provide a simple and robust environment which enables users and application developers to produce applications that can exploit the full power and possibilities of the grid.

GridLab has a layered architecture which consists of application, GAT, and service layer. In application layer, the applications can access all capability providers they need via the grid application toolkit (GAT) API. The GAT which is the main deliverable of the GridLab project achieves the interaction with all external capability providers in the behavior of user applications. Hence, GAT provides application programmers with a single interface to a grid environment. In service layer of GridLab, services are designed to complement and complete the existing grid infrastructure, and to provide functionality needed by the GridLab applications. The main services which have been implemented in this layer are: GRMS, GAS, data movement service, and third party services.

GridLab aims to provide an environment that allows application developers to use the grid without having to understand, or even being aware of the underlying technologies. Lots of application frameworks (e.g., Cactus [Goodale et al., 2003] and Triana [Churches et al., 2005]) are built upon GridLab and help prototype the GAT interface.

1.6.3.2   Distributed interactive engineering toolbox

Distributed Interactive Engineering Toolbox (DIET) is a hierarchical set of components used for the development of applications based on computational servers on the grid. It consists of a set of elements that can be used together to build applications using the GridRPC paradigm [Amar et al., 2006].

The DIET has a hierarchical architecture providing flexibility and can be adapted to various environments including heterogeneous network hierarchies. In DIET, there are three main components: master agent (MA), local agent (LA), and server daemon (SeD). (1) *Master agent* is the entry point of DIET system. Clients submit requests for a specific computational service to the MA. The MA then forwards the request in the DIET hierarchy until the request reaches the SeDs. LA aims at transmitting requests and information between MAs and servers. (2) *Local agent* maintains the information about the list of requests and, for each of its subtrees, the number of servers that can solve a given problem. (3) *Server daemon* provides the interface to computational servers and can offer any number of application specific computational

services. The information stored on a SeD is the list of the data available on
a server, the list of problems that can be solved on it, and every information
concerning its load (CPU capacity, available memory, etc.).

The management of the platform is handled by several tools like GoDIET
for the automatic deployment of the different components, LogService for
monitoring, and VizDIET for the visualization of the behavior of DIET's
internals. DIET provides a special feature for scheduling through its plug-
in schedulers. The DIET user is provided with the possibility of defining
requirements for scheduling of tasks by configuring the appropriate scheduler.

1.6.3.3 XtremWeb

XtremWeb is a global computing system which achieves the secure and fault
tolerant peer to peer computing by remote procedures call (RPC) technology
[Cappello et al., 2005]. XtremWeb allows clients to submit task requests to
the system which will execute them on workers. In order to decouple clients
from workers and to coordinate task executions on workers, the coordinator
is added between client and worker nodes. The details of these three services
are shown as follows:

- Coordinator: The coordinator in XtremWeb is composed of three services: the repository, the scheduler, and the result server. The coordinator accepts task requests coming from clients, assigns the tasks to
  the workers according to a scheduling policy, supervises task execution
  on workers, detects works crash, reschedules crashed tasks on any other
  available worker, and delivers task results to client upon request.

- Worker: The worker architecture includes four components: the task
  pool, the execution thread, the communication manager and the activity
  monitor. The activity monitor detects the host information (e.g., per-
  centage of CPU idle, mouse and keyboard activity) to control whether
  computations can be started on the hosting machine. The task pool is a
  queue structure of tasks, maintained by scheduling strategy. The communication manager ensures communications with other entities and
  achieves files downloading and results uploading. The execution thread
  extracts task from the task pool, starts computation, and waits for the
  task to complete.

- Client: The client in XtremWeb is implemented as a library plus a
  daemon process. The library provides an interface which can be used
  to achieve the interaction between the application and the coordinator.
  The daemon process makes recovery points regularly which insure the
  machine recovery and jobs rescheduling on another machine.

XtremWeb aims to turn a large scale distributed system into a parallel computer with classical users, administration, and programming interface using
fully decentralized mechanism to implement the system functionality.

1.6.4   Grid programming environments

1.6.4.1   Cactus code

Cactus is a framework for building a variety of computing applications in science and engineering, including astrophysics, relativity and chemical engineering [Goodale et al., 2003]. Cactus which started in 1997 is an open source problem solving environment designed for scientists and engineers. Its modular structure easily enables parallel computation across different architectures and collaborative code development between different groups. Cactus originated in the academic research community, where it was developed and used over many years by a large international collaboration of physicists and computational scientists.

Cactus consists of a central core (or "flesh") and application modules (or "thorns"). A thorn is the basic working module within Cactus. All user-supplied code goes into thorns, which are independent of each other and provide a range of computational capabilities, such as parallel I/O, data distribution, or checkpointing. The flesh is independent of all thorns and provides the main program, which parses the parameters and activates the appropriate thorns, passing control to thorns as required. The flesh connects to thorns through an extensible interface.

As a portable system, Cactus runs on many architectures. Applications, developed on standard workstations or laptops, can be seamlessly run on clusters or supercomputers. Cactus provides easy access to many cutting edge software technologies being developed in the academic research community, including the Globus Metacomputing Toolkit, HDF5 parallel file I/O, the PETSc scientific library, adaptive mesh refinement, web interfaces, and advanced visualization tools.

1.6.4.2   GrADS

The goal of the grid application development software (GrADS) project is to simplify distributed heterogeneous applications development and to make it easier for ordinary scientific users to develop, execute, and tune applications on the grid [Berman et al., 2001].

There are two sub-systems in a GrADS environment: GrADS program preparation system and GrADS execution environment.

- GrADS program preparation system: In order to simplify development of grid-enabled applications, the preparation system provides a methodology in which most users will compose applications in a high-level, domain-specific language built upon pre-existing component libraries. This approach hides grid-level details and lets the user express a computation in terms that make sense to an expert in the application domain. Two components are implemented to support this methodology: a domain-specific language called telescoping language, and a tool for load-time tailoring which is called the dynamic optimizer. The SaNS libraries which automatically select and integrate the most effective library components for a given problem, data set, and collection of resources are developed to improve the ability of computational scientists to solve challenging problems efficiently.

- GrADSoft execution environment: The GrADSoft execution environment provides new mechanism to discover and communicate information about the environment to program components, to transfer program requirements to the environment and program components in ways of which admit to effective control, and to monitor and control adaptively an executing program.

The GrADS project has established an effort to pioneer technologies that will be needed for ordinary scientific users to develop applications for the grid. These technologies include a new program preparation framework and an execution environment that employs continuous monitoring to ensure that reasonable progress is being made toward completion of a computation.

1.6.4.3 CoG kits

There are two important concepts in the distributed computing world which have evolved in parallel: "commodity" and "grid" computing [von Laszewski et al., 2000]. The commodity computing concerns a broad spectrum of distributed computing technologies (i.e., web protocols, Java, JINI, CORBA, DCOM, etc.) which has emerged with revolutionary effects on how we access and process information. Nevertheless, grid computing focus on the coordinated use of distributed high-end resources for scientific problem solving, specially in the high-performance computing community. In order to enable developers of grid applications to exploit commodity technologies and to export grid technologies to commodity computing, the combination of the worlds of commodity and grid computing creates thus the CoG kits which is a set of general components that map grid functionality into specific commodity environments or frameworks.

The mapping of grid and commodity technologies is not simply an interface definition problem. The expression of grid concepts and services in a particular commodity framework is the important issue which must be solved. The Figure 1.3 illustrates the mapping schema. The requirements of the science portals and other applications have motivated the CoG kit developers to explore mappings to several languages. Perl and Python are explored to support web-based programming based on CGI scripts; in order to support graphical user interface development and the ability to run many grid services through Java-enabled web browsers, CoG kits provide Java mapping. The Common object request broker architecture (CORBA) and the distributed component object model (DCOM) are considered to address the issue of accessing grid services through high-level distributed computing frameworks.

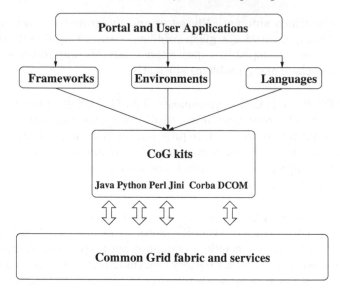

FIGURE 1.3: The schema illustrating the mapping of grid functionality and specific commodity environments with CoG kits.

The Java commodity grid toolkit (CoG kit) is the first attempt of CoG project. In Java CoG kit, a rich set of classes that provide the Java programmer with access to basic grid services, enhanced services suitable for the definition of desktop problem solving environments, and a range of GUI elements have been defined. It has proved possible to recast major Grid services in Java terms without compromising on functionality.

1.6.5   Grid portals

One of the areas of grid application that are focused on at this time is the development of gateways and grid portals, which is a web-based single point of entry to a grid and its implemented services. With the widespread development of the Internet, scientists expect to expose their data and applications through portals. The grid portals provide user-friendly web page interfaces facilitating grid applications users to perform operations on the grid and access grid resources specific to a particular domain of interest.

Currently, there are various technologies and toolkit that can be used for grid portal development. According to [Yang et al., 2006], grid portals can be classified into nonportlet-based and portlet-based.

- Nonportlet-based portal is a grid portal that is designed based on typical three-layers architecture. The first layer is the user layer, which aims to provide the user-friendly interface for user. User layer is responsible for displaying the portal content; it can be web browser, or other desktop tools. The second layer is the grid service layer, including authentication service, job management service, information service, file service, security service. The authentication service allows portal to authenticate users. Once authenticated, users can use other services to access resources of the system (e.g., job management service for submitting jobs on a remote machine, information service for monitoring jobs submitted, and viewing results). The second layer receives HTTP requests from the first layer and interacts with the third layer for performing the grid operations on relevant grid resource and retrieving the executed result from grid resources. The third layer is a backend resource layer, which consists of computation, data and application resources.

- Portlet-based portal includes a collection of portlets. A portlet is a web component that generates fragments – pieces of markup (e.g., HTML, XML) adhering to certain specifications (e.g., JSR-168 [Sun, 2009a], WSRP [OASIS, 2009]). Portlets improve the modular flexibility of developing grid portals as they are pluggable and can be aggregated to form a complete web page depending on user needs.

1.6.5.1 P-GRADE portal

P-GRADE grid portal [Kacsuk et al., 2006] is the first grid portal that tries to solve the interoperability problem at the workflow level with great success. It is a workflow-oriented grid portal with the main goal to support all stages of grid workflow development and execution processes.

The P-GRADE portal provides the following functions (see Figure 1.4): communicating with the portal server, users can achieve the functions of defining grid environments, managing grid certificates, controlling the execution of workflow applications, and visualizing the progress of workflows; workflow editor can perform the creation and modification of workflow applications [Kertesz et al., 2006].

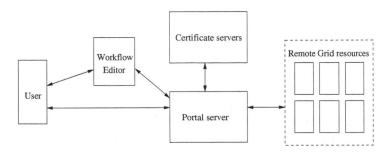

FIGURE 1.4: P-GRADE portal system functions.

During the workflow editing the user has the possibility to select a grid resource for each job, or let a broker choose one. Currently there are two brokers used by the portal: the LCG-2 broker and GTbroker. The GTbroker interact with the Globus resources to perform job submission. The static and dynamic information of grid resources are collected by GTbroker to achieve scheduling activities. The LCG-2 broking solution is used to reach LCG-2 based grids. The mission of the LHC computing project (LCG) is to build and maintain a data storage and analysis infrastructure for the entire high energy physics community that will use the LHC. The Large Hadron Collider (LHC), built at CERN near Geneva, is the largest scientific instrument on the planet and it begins operations in 2007. With exploiting the broking functions of GTbroker and LCG-2 broker, users can develop and execute multi-grid workflows in a convenient environment.

The integration of P-GRADE into GEMLCA shows the use of portal in a grid environment [Kacsuk et al., 2006]. Grid execution management for legacy code applications (GEMLCA) represents a general architecture for deploying legacy applications as grid services without re-engineering the code or even requiring access to the source files. GEMLCA adds an additional layer to wrap the legacy application on top of a service-oriented grid middleware, like Globus Toolkit version 4 (GT4). GEMLCA communicates with the client through SOAP-XML messages, gets input parameter values, submits the legacy executable to a local job manager like Condor or portable batch system (PBS), and returns the results to the client in SOAP-XML format. GEMLCA provides the capability to convert legacy codes into grid services. However, an end-user without specialist computing skills still requires a user-friendly web interface (portal) to access the GEMLCA functionalities. In order to solve this problem, GEMLCA is integrated with the P-GRADE grid portal. Following this integration, legacy code services can be included in end-user workflows, running on different GEMLCA grid resources. The workflow manager of the portal contacts the selected GEMLCA resource and passes the actual parameter values of the legacy code to it. Then the GEMLCA resource executes the legacy code with these actual parameter values and delivers the results back to the portal.

1.6.5.2   GridSphere

GridSphere [GridSphere, 2009] is a typical portlet-based portal. The Grid-Sphere portal framework is developed as a key part of the European project GridLab [GridLab, 2009]. It provides an open-source portlet based web portal and enables developers to quickly develop and package third-party portlet web applications that can be run and administered within the GridSphere portlet container. Two key features of GridSphere framework are: (i) allowing administrators and individual users to dynamically configure the content based on their requirements, and (ii) supporting grid-specific portlets and APIs for grid-enabled portal development. However, the main disadvantage of the current version of GridSphere (i.e., GridSphere 2.1) is that it does not support WSRP specification.

1.6.5.3 Other portal systems

The Pegasus [Singh et al., 2005] portal provides an HTTP(S)-based interface that can be accessed using a standard web browser. The portal architecture is composed of three layers. The top layer consists of the user machines and web browsers. The second layer consists of the web application server hosting the portal. The server is multithreaded and can handle multiple user requests at the same time. The third layer consists of the grid components and services used by the portal.

In order to use the Pegasus grid portal the user needs to have a valid grid credential in a MyProxy server. The portal does not provide access to a predetermined set of resources. Instead, the user can specify the resources to be used. From the web browser, the users specify the parameters of the application and Pegasus does the mapping of tasks in the workflow to resources specified in the resource configuration. The submitted workflow may take a long time to complete. The user may logout from the portal and login later to check its status. The portal allows also users to view the status of the workflow (submitted, active, done, failed), the number of tasks already completed, the tasks currently executing, and other information.

The Pegasus grid portal is very useful in scenarios where a virtual organization (VO) wants to provide easy to use application submission interface to its members. It is able to map abstract workflow onto physical resources; thus users are shielded from the complexity of installing and using the various components in order to access the Grid resources.

GridFlow [Cao et al., 2003] is a grid workflow management system developed at the University of Warwick. Rather than focusing on workflow specification and the communication protocol, GridFlow is more concerned about service-level scheduling and workflow management. The GridFlow portal performs two level functions: global grid workflow management and local grid sub-workflow scheduling. The execution and monitoring functionalities are provided at the global grid level, which work on top of an existing agent-based grid resource management system. At each local grid, sub-workflow scheduling and conflict management are processed on top of an existing performance prediction based task scheduling system. A fuzzy timing technique is applied to address new challenges of workflow management in a cross-domain and highly dynamic grid environment.

1.7   Grid evolution

As soon as computers are interconnected and communicate, the research in designing, building, and deploying distributed computer systems was begun to explore. An increasing number of research groups have been working in the field of wide-area distributed computing. Middleware, libraries and tools have been implemented to coordinate geographically distributed resources for the execution of a range of parallel and distributed applications. With the technologies innovation, the distributed computing has been known sequentially by several names, such as metacomputing, scalable computing, global computing, Internet computing, and more recently grid computing.

According to the paper [de Roure et al., 2003], the grid computing can be identified by three stages of grid evolution:

The first generation: The first generation of grid is known as metacomputing which emerged in mid 1990s. The objective of these early metacomputing projects was to provide computational resources to a range of high performance applications. Two representative projects are FAFNER [de Roure et al., 2003] and I-WAY [Foster et al., 1997]. FAFNER is a ubiquitous system that worked on any platform with a web server. Typically the applications in FAFNER can be separated into independent sections, which are executed in parallel in each computing resource without the need of a fast interconnect. Its clients are low end computers and it depends on a lot of human intervention to distribute and collect computing results. I-WAY is designed to cope with a range of diverse high performance applications that typically needed a fast interconnect and powerful resources at multiple supercomputing centers. The experiences and software developed as part of the I-WAY project have been fed into the Globus project.

The second generation: In the second generation, the core software for the grid has evolved from providing dedicated services for large and computationally intensive high performance computing, to the more generic and open deployment of middleware. Based on this core software, a range of accompanying tools and utilities are developed, providing higher-level services to both users and applications, and spanning resource schedulers and brokers as well as domain specific users interfaces and portals. These projects and utilities that emerged in the second generation have been discussed in Section 1.6. Peer-to-peer techniques have also emerged during this period.

Service-oriented computing (the third generation): New grid applications need to be able to reuse existing components and information resources, and to assemble these components in a flexible manner. It was apparent that the service-oriented architecture provided the flexibility required for the third generation grid. The open grid services architecture (OGSA) framework is the convergence of web services and grid computing, and it supports the creation, maintenance, and application of ensembles of services maintained by virtual organizations (VOs). The services here are more standard, easily interactive, and metadata-enabled; thus they are more adopted in the e-science infrastructure.

The evolution of the grid is a continuous process. The technologies of semantic web and workflow which we will discuss in the next chapters have been more integrated into the grid and grid services. Perhaps, a semantic grid with workflow creation and scheduling capacities will be the next generation of grid.

1.8 Concluding remarks

In this chapter, the concept of grid is first introduced. With about a decade of development, lots of grid infrastructures and utilities have emerged and the manner and domain of a grid is now employed have varied vastly, spanning from high-throughput computing to on-demand computing, from scientific research to e-business. The grid evolves continuously and the third generation of grid, the service oriented computing, has emerged. The technologies of semantic web and workflow have been used in the grid and the integration and convergence of technologies make the grid to provide more flexible, automatic and complex grid services to fulfill industrial and commercial needs.

1.9 References

[Allen et al., 2003] Allen, G., Goodale, T., Radke, T., Russell, M., Seidel, E., Davis, K., Dolkas, K. N., Doulamis, N. D., Kielmann, T., Merzky, A., Nabrzyski, J., Pukacki, J., Shalf, J., and Taylor, I. (2003). Enabling applications on the grid: a Gridlab overview. *International Journal of High Performance Computing Applications*, 17:449–466.

[Amar et al., 2006] Amar, A., Bolze, R., Bouteiller, A., Chouhan, P. K., Chis, A., Caniou, Y., Caron, E., Dail, H., Depardon, B., Desprez, F., Gay, J.-S., Mahec, G. L., and Su, A. (2006). DIET: new developments and recent results. In *Proceedings of CoreGRID Workshop on Grid Middleware (in conjunction with EuroPar2006)*, Dresden, Germany.

[Berman et al., 2001] Berman, F., Chien, A., Cooper, K., Dongarra, J., Foster, I., Gannon, D., Johnsson, L., Kennedy, K., Kesselman, C., Mellor-Crummey, J., Reed, D., Torczon, L., and Wolski, R. (2001). The GrADS project: software support for high-level grid application development. *International Journal of High Performance Computing Applications*, 15(4):327–344.

[Bode et al., 2000] Bode, B., Halstead, D. M., Kendall, R., Lei, Z., and Jackson, D. (2000). The portable batch scheduler and the Maui scheduler on Linux clusters. In *Proceedings of the 4th Conference on Linux Showcase (ALS'00)*, pages 27–27, Berkeley, CA, USA. USENIX Association.

[Bote-Lorenzo et al., 2003] Bote-Lorenzo, M. L., Dimitriadis, Y. A., and Gómez-Sánchez, E. (2003). Grid characteristics and uses: a grid definition. In *Proceedings of the 1st European Across Grids Conference*, Lecture Notes in Computer Sciences, pages 291–298, Santiago de Compostela, Spain. Springer-Verlag.

[Cao et al., 2003] Cao, J., Jarvis, S. A., Saini, S., and Nudd, G. R. (2003). GridFlow: workflow management for grid computing. In *Proceedings of the 3rd International Symposium on Cluster Computing and the Grid (CCGRID'03)*, page 198. IEEE Computer Society.

[Cappello et al., 2005] Cappello, F., Djilali, S., Fedak, G., Herault, T., Magniette, F., Néri, V., and Lodygensky, O. (2005). Computing on large-scale distributed systems: XtremWeb architecture, programming models, security, tests and convergence with grid. *Future Generation Computer Systems*, 21(3):417–437.

[Churches et al., 2005] Churches, D., Gombas, G., Harrison, A., Maassen, J., Robinson, C., Shields, M., Taylor, I., and Wang, I. (2005). Programming scientific and distributed workflow with Triana services. *Concurrency: Practice and Experience.*

[Condor, 2009] Condor (2009). Documentation. Available online at: `http://www.cs.wisc.edu/condor` (accessed May 1, 2009).

[de Roure et al., 2003] de Roure, D., Baker, M. A., Jennings, N. R., and R.Shadbolt, N. (2003). *Grid computing: making the global infrastructure a reality*, chapter The evolution of the grid, pages 65–100. John Wiley & Sons Ltd., New York.

[Foster, 2002] Foster, I. (2002). What is the grid? A three point checklist. *GRIDtoday*, 1(6). Available online at: `http://www.gridtoday.com/02/0722/100136.html` (accessed May 1, 2009).

[Foster, 2006] Foster, I. (2006). Globus toolkit version 4: software for service-oriented systems. *Journal Comput. Sci. Technol.*, 21(4):513–520.

[Foster et al., 1997] Foster, I., Geisler, J., Nickless, W., Smith, W., and Tuecke, S. (1997). Software infrastructure for the I-WAY high performance distributed computing experiment. In *Proceedings of the 5th IEEE Symposium on High Performance Distributed Computing*, pages 562–571.

[Foster and Kesselman, 1997] Foster, I. and Kesselman, C. (1997). Globus: a metacomputing infrastructure toolkit. *The International Journal of Supercomputer Applications and High Performance Computing*, 11(2):115–128.

[Foster and Kesselman, 1998] Foster, I. and Kesselman, C. (1998). *The grid: blueprint for a new computing infrastructure*. Morgan Kaufmann Publishers Inc., San Francisco, CA, USA. Available online at: `http://portal.acm.org/citation.cfm?id=289914` (accessed May 1, 2009).

[Foster and Kesselman, 1999] Foster, I. and Kesselman, C. (1999). *The grid: blueprint for a new computing infrastructure*, chapter Computational grids, pages 15–51. Morgan Kaufmann Publishers Inc., San Francisco, CA, USA.

[Foster et al., 2002] Foster, I., Kesselman, C., Nick, J. M., and Tuecke, S. (2002). Grid services for distributed system integration. *Computer*, 35(6):37–46.

[Foster et al., 2001] Foster, I., Kesselman, C., and Tuecke, S. (2001). The anatomy of the grid: enabling scalable virtual organizations. *International Journal of High Performance Computing Applications*, 15(3):200–222.

[Frey et al., 2002] Frey, J., Tannenbaum, T., Livny, M., Foster, I., and Tuecke, S. (2002). Condor-G: a computation management agent for multi-institutional grids. *Cluster Computing*, 5(3):237–246.

[Goodale et al., 2003] Goodale, T., Allen, G., Lanfermann, G., Massó, J., Radke, T., Seidel, E., and Shalf, J. (2003). The Cactus framework and toolkit: design and applications. In *Proceedings of the 5th International Conference of Vector and Parallel Processing (VECPAR'2002)*, Lecture Notes in Computer Sciences. Springer-Verlag.

[GridLab, 2009] GridLab (2009). Documentation. Available online at: `http://www.gridlab.org` (accessed May 1, 2009).

[GridSphere, 2009] GridSphere (2009). Documentation. Available online at: `http://www.gridsphere.org` (accessed May 1, 2009).

[Kacsuk et al., 2006] Kacsuk, P., Kiss, T., and Sipos, G. (2006). Solving the grid interoperability problem by P-GRADE portal at workflow level. In *Proceedings of the 15th International Symposium on High Performance Distributed Computing (HPDC-15)*. IEEE Computer Society.

[Kertesz et al., 2006] Kertesz, A., Farkas, Z., Kacsuk, P., and Kiss, T. (2006). Multiple broker support by grid portals. In *Proceedings of the CoreGRID Workshop on Grid Middleware*.

[Krauter et al., 2002] Krauter, K., Buyya, R., and Maheswaran, M. (2002). A taxonomy and survey of grid resource management systems for distributed computing. *International Journal of Software Practice and Experience*, 32(2):135–164.

[Nakada et al., 2007] Nakada, H., Sato, H., Saga, K., Hatanaka, M., Saeki, Y., and Matsuoka, S. (2007). Job invocation interoperability between NAREGI middleware beta and gLite. In *Proceedings of HPC Asia 2007*, pages 151–158.

[Natrajan et al., 2001] Natrajan, A., Humphrey, M., and Grimshaw, A. (2001). Grids: harnessing geographically-separated resources in a multi-organisational context. In *Proceedings of the 15th Annual International Symposium on High Performance Computing Systems and Applications*.

[OASIS, 2009] OASIS (2009). WSRP: web services for remote portlets. Available online at: `http://www.oasisopen.org/committees/tc_home.php?wg_abbrev=wsrp` (accessed May 1, 2009).

[PBS, 2006] PBS (2006). Portable batch system (PBS). Available online at: `http://www.openpbs.org/` (accessed May 1, 2009).

[Platform, 2009] Platform (2009). Lsf. Available online at: `http://www.platform.com/Products/Platform.LSF.Family/` (accessed May 1, 2009).

[Raman et al., 1998] Raman, R., Livny, M., and Solomon, M. (1998). Matchmaking: distributed resource management for high throughput computing. In *Proceedings of the 7th HPDC*.

[Singh et al., 2005] Singh, G., Deelman, E., Mehta, G., Vahi, K., Su, M., Berriman, B., Good, J., Jacob, J., Katz, D., Lazzarini, A., Blackburn, K., and Koranda, S. (2005). The Pegasus portal: web based grid computing. In *Proceedings of the 20th Annual ACM Symposium on Applied Computing*.

[Streit et al., 2005] Streit, A., Erwin, D., Lippert, T., Mallmann, D., Menday, R., Rambadt, M., Riedel, M., Romberg, M., Schuller, B., and Wieder, P. (2005). UNICORE: from project results to production grids. In *Grid computing: the new frontier of high performance computing*, volume 14 of *Advances in Parallel Computing*, pages 357–376, Amsterdam, The Netherlands. Elsevier Science Publishers B.V.

[Sun, 2009a] Sun (2009a). Introduction to JSR-168. Available online at: `http://developers.sun.com/prodtech/portalserver/reference/techart/jsr168/` (accessed May 1, 2009).

[Sun, 2009b] Sun (2009b). Sun grid engine. Available online at: `http://gridengine.sunsource.net` (accessed May 1, 2009).

[Tannenbaum et al., 2001] Tannenbaum, T., Wright, D., Miller, K., and Livny, M. (2001). Condor: a distributed job scheduler. In Sterling, T., editor, *Beowulf Cluster Computing with Linux*. MIT Press.

[Thain et al., 2005] Thain, D., Tannenbaum, T., and Livny, M. (2005). Distributed computing in practice: the Condor experience. *Concurrency: Practice and Experience*, 17(2 4):323 356.

[von Laszewski et al., 2000] von Laszewski, G., Foster, I., and Gawor, J. (2000). CoG kits: a bridge between commodity distributed computing and high-performance grids. In *Proceedings of the ACM 2000 Conference on Java Grande (JAVA '00)*, pages 97–106, New York, NY, USA. ACM Press.

[Yang et al., 2006] Yang, X., Dove, M. T., Hayes, M., Calleja, M., He, L., and Murray-Rust, P. (2006). Survey of major tools and technologies for grid-enabled portal development. In *Proceedings of the UK e-Science All Hands Meeting 2006*, Nottingham, UK.

Chapter 2

Synchronization protocols for sharing resources in grid environments

Julien Sopena

LIP6/INRIA Regal Team, Université Pierre et Marie Curie, 104 avenue du Président Kennedy, 75016 Paris, France

Luciana Arantes

LIP6/INRIA Regal Team, Université Pierre et Marie Curie, 104 avenue du Président Kennedy, 75016 Paris, France

Fabrice Legond-Aubry

LIP6/INRIA Regal Team, Université Pierre et Marie Curie, 104 avenue du Président Kennedy, 75016 Paris, France

Pierre Sens

LIP6/INRIA Regal Team, Université Pierre et Marie Curie, 104 avenue du Président Kennedy, 75016 Paris, France

2.1 Introduction

Grids are extremely interesting for executing distributed and/or parallel applications that require a lot of computational power, data storage or access to resources that are not available locally. Since these applications share grid resources, the latter should be accessed by the application processes in an exclusive way, i.e., exactly one process can access the shared resource at any given time (*safety* property) and all access requests are eventually satisfied (*liveness* property). A process's segment of code that accesses a shared resource is called a critical section (*CS*). Therefore, a synchronization protocol that provides mutual exclusion is extremely important for such applications. However, a Grid platform is usually composed of a large number of clusters. As such clusters are usually spread out over different sites, cities or even countries, communication in a Grid environment is intrinsically heterogeneous. Nodes within one cluster are linked by local networks (LAN) whereas clusters are connected by wide area network (WAN) links. Grids present thus a hierarchy of communication delays where the cost of sending a message between nodes of different clusters is much higher than sending the same message between nodes within the same cluster. Hence, a synchronization protocol that offers mutual exclusion must be tailored to the latency hierarchy of the Grid for performance reason since the performance of the mutual exclusion protocol can have a major impact on the overall performance of applications.

Distributed mutual exclusion algorithms can be divided into two families: *permission-based* (e.g., Lamport [Lamport, 1978], Ricart-Agrawala [Ricart and Agrawala, 1981], Maekawa [Maekawa, 1985]) and *token-based* (Suzuki-Kazami [Suzuki and Kasami, 1985], Raymond [Raymond, 1989], Naimi-Tréhel [Naimi and Tréhel, 1996], Martin [Martin, 1985]). The algorithms of the first family are based on the principle that a node enters a critical section only after having received permission from all the other nodes (or a majority of them [Maekawa, 1985]). In the second group of algorithms, a system-wide unique token is shared among all nodes, and its possession gives a node the exclusive right to execute a critical section. Token-based algorithms present different solutions for the transmission and control of critical section requests of processes. Each solution is usually expressed by a logical topology that defines the paths followed by critical section request messages which might be completely different from the physical network topology.

With regard to the number of nodes, token-based mutual exclusion algorithms present an average message traffic which is lower than that of permission-based ones. Thus, they are more suitable for controlling concurrent access to shared resources of Grids whose number of nodes is often very large. However, existing token-based algorithms still have intrinsic limits and do not take into account the above-mentioned hierarchy of communication latencies. Our proposal is a generic composition approach which enables the combination of any two token-based mutual exclusion algorithms: one at *intra-cluster* level and a second one at *inter-cluster*. Hence, by using our composition mechanism, synchronization protocols that ensure mutual exclusion for Grid application can be easily deployed by just "plugging in" token-based algorithms on each levels of the hierarchy. The choice of such algorithms thus takes into account communication latency heterogeneity of Grids. Furthermore, the extensive performance evaluation tests that we have conducted on

Grid'5000 [1] show that the good choice for an *inter-cluster* mutual exclusion algorithm depends on the application behavior, i.e., the frequency with which the application processes request for the shared resource. In other words, the parallelism degree of the application has an impact on the choice of *inter-cluster* algorithms. Therefore, relying on both the application behavior and our performance study results, effective synchronization protocols that provide token-based mutual exclusion tailored to a specific Grid application can be built on top of our compositional approach.

2.2   Token-based mutual exclusion algorithms

Several token-based mutual exclusion algorithms have been proposed in the literature. This section presents three of them, **Martin**'s [Martin, 1985], **Naimi-Tréhel**'s [Naimi and Tréhel, 1996], and **Suzuki-Kasami**'s [Suzuki and Kasami, 1985], since we used them in our performance experiments, described in Section 2.6.

Martin's, **Naimi-Tréhel**'s, and **Suzuki-Kasami**'s algorithms are respectively based on a *ring*, a *tree*, and a *complete logical connection graph* for forwarding critical section requests. As they present distinct solutions for both transmitting requests and controlling the algorithm's liveness, they present different message complexity with regard to the number of nodes.

2.2.1   Martin's algorithm

Martin's algorithm considers that nodes are organized in a logical ring. Requests for the token move along one direction while the token move on the opposite direction, as shown in Figures 2.1(b) and 2.1(c).

When node i, which does not hold the token, wants to enter the critical section (CS) it asks for the token by sending a request message to its successor j in the ring. If j does not keep the token it forwards the request to its successor. The request will travel along the ring till it reaches the site k which keeps the token. Upon receiving the request, if k is not in CS itself, it forwards the token to its predecessor. Each node between k and i will do the same. Therefore, the token will eventually reach i, which can then enter the CS. Notice that before the token reaches i, nodes between i and k might have requested the token too. Thus, when k forwards the token on behalf of i all

---

[1]Grid'5000 is an initiative from the French Ministry of Research through the ACI GRID incentive action, INRIA, CNRS and RENATER and other contributing partners (see https://www.grid5000.fr).

pending requests of nodes between k and i will be satisfied when they receive the token.

Notice that for optimization reasons, upon receiving a request from its predecessor, a node that is also requesting the token does not need to forward the request of the predecessor. It just keeps the information that after satisfying its own request, it must send the token to its predecessor.

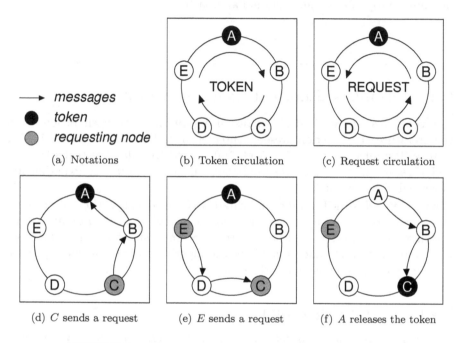

FIGURE 2.1: Execution example of Martin's algorithm.

Figures 2.1(d), 2.1(e) and 2.1(f) show the execution of Martin's algorithm for 5 nodes (A, B, C, D, and E). Initially A keeps the token. Since node C wants to enter the critical section, it sends a request to its neighbor B (Figure 2.1(d)). Upon receiving C's request, B registers that there is a pending request and forwards it to A. In Figure 2.1(e), E also decides to ask for the token by sending a request to D. Similarly to B, D registers the request and sends it to C, which has already a registered pending request. Thus, C will not forward the request to B.

When A releases the critical section (Figure 2.1(f)), it sends the token to B which forwards it to C, which has asked for it. Afterwards, when C releases the critical section, it will forward the token to D which then will send it to E that enters the critical section too.

2.2.2 Naimi-Tréhel's algorithm

Naimi-Tréhel algorithm [Naimi and Tréhel, 1996] maintains a logical dynamic tree structure, named the *LAST tree*, such that the root of the tree is always the last node that will receive the token among current requesting nodes. A second structure, named the *NEXT queue*, is a distributed queue that controls the nodes which are waiting for the token. Each node just holds information about its *LAST* and *NEXT* nodes.

Initially, the root of the *LAST* tree is the token holder and the *LAST* of all other nodes points to the root. A token request travels along a path of *LASTs* to the root. When receiving this request, each node along this path sets its *LAST* pointer to the current requester, i.e., the tree is modified dynamically. When a request arrives at the root, the latter updates its *NEXT* to point to the requester. When a site releases the CS, the token is sent to the site indicated by its *NEXT* pointer.

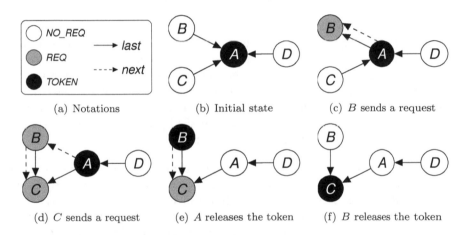

(a) Notations (b) Initial state (c) *B* sends a request

(d) *C* sends a request (e) *A* releases the token (f) *B* releases the token

FIGURE 2.2: Execution example of Naimi-Tréhel's algorithm.

Figure 2.2 shows an example of a Naimi-Tréhel execution with 4 nodes. Initially, site *A* is the root and holds the token (Figure 2.2(b)) and the *LAST* of all other nodes point to *A*. In Figure 2.2(c), node *B* asks for the token by sending a request to its *LAST* ($LAST_B = A$). *B* becomes the new root ($LAST_B = NIL$). Upon receiving *B*'s request, *A* updates its *NEXT* and *LAST* variables to point to *B*. In Figure 2.2(d), *C* asks *A* for the token. The request is forwarded to *B* which updates its *NEXT* to *C* ($NEXT_B = C$). Both *A* and *B* update their *LAST* to *C*, since the latter is the last requester of the token (*C* becomes the new root of the *LAST* tree). When *A* releases the critical section, the token will be sent to *B* as $NEXT_A = B$, as shown in

Figure 2.2(e). Similarly, when finishing executing the critical section, B sends the token to its $NEXT$ which is C (Figure 2.2(f)).

2.2.3　Suzuki-Kasami's algorithm

In Suzuki-Kazami's algorithm, when a node i, which does not hold the token, attempts to enter the critical section, it diffuses a request message to the other $N - 1$ nodes. Such a message contains the identifier i of the node and a sequence number x which indicates the xth critical section invocation of i. As in the previous token-based algorithms, when node i receives the token, it enters the critical section.

Each node i keeps an array RN_i of size N where it stores the largest token invocation (sequence number) of each node of which it is aware. Whenever i receives a request from j, it updates $RN_i[j]$ with the sequence number of the request.

The *token* message includes a queue Q of nodes whose requests are pending and an array LR of size N which keeps the sequence number of the most recent satisfied request from each node. When node i exits the critical section, it updates $LR[i]$ with its current $RN_i[i]$ in order to inform that its last request has been satisfied. Then, it appends to Q all nodes not yet in Q for which it knows that their requests have not been satisfied yet. If Q is not empty, the first node j is removed from Q and the token is sent to j.

Figure 2.3 shows an execution of the algorithm. We consider that node A has already executed 6 critical sections ($LR[A] = 6$) and it is currently in its 7th critical section ($RN_A[A] = 7$) as shown in Figure 2.3(b). Suppose then that node C decides to ask for the critical section (Figure 2.3(c)). To this end, C increases the entry of its RN that corresponds to itself ($RN_C[C] = 9$) and broadcasts its request with such a value for the token to the other nodes. Upon reception of the request, node i updates its $RN_i[C]$. Let also suppose that node D, Figure 2.3(d), requests the token which corresponds to its third request $RN_D[D] = 3$.

Upon releasing the critical section, node A registers it in LR vector ($LR[A] = 7$). It then compares its RN_A vector with LR and it includes in Q all the nodes whose entries in RN are greater than LR which are not already in Q, i.e., the pending requests. In this case, nodes C and D are included in Q (Figure 2.3(e)). Node A then removes the first node in Q, which is node C, and sends the token to it. In its turn, when node C finishes executing the CS, it compares the two vectors. There is still a difference which concerns node D, but which is already in Q. Thus, it removes D from Q and sends the token to it (Figure 2.3(f)).

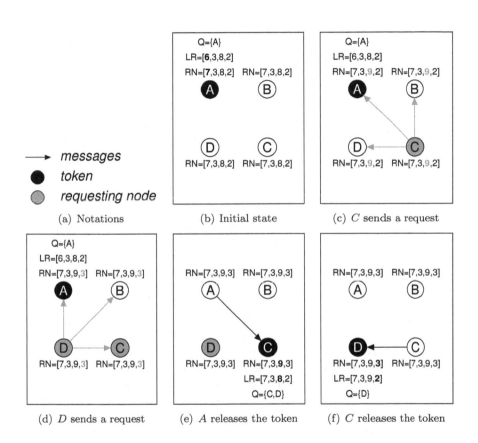

FIGURE 2.3: Execution example of Suzuki-Kasami's algorithm.

2.3   Mutual exclusion algorithms for large configurations

Several works found in the literature have proposed to adapt or compose existing mutual exclusion algorithms. By exploiting the logical or physical topological architectures of clusters (groups), they aim at reducing the number of messages or the waiting time delay to enter the CS. Basically, they adopt one of the two following approaches: in the first one [Bertier et al., 2006], [Mueller, 1998], the order of the pending requests of the waiting queue is changed in order to satisfy requests with higher priority first. Thus, priority of requests can dynamically change based on their locality which result in satisfying requests from the same cluster before those from distant ones. In the second approach, authors compose mutual exclusion algorithms and map them to the hierarchy of the architecture (physical or logical): one at the intra-cluster or intra-group level and a second one at the inter-cluster or inter-group level.

2.3.1   Priority-based approach

Bertier et al. [Bertier et al., 2004] adapt Naimi-Tréhel's algorithm by changing the order of the pending requests of the $NEXT$ queue in order to satisfy intra-cluster requests before inter-cluster ones. Local cluster preemptions of the token take place since higher priority is given to requests of local cluster nodes. A threshold for the maximum number of per cluster preemption which characterizes the degree of locality of the cluster is defined for avoiding starvation. While the number of local requests is below such a threshold, the $NEXT$ queue is modified in order to satisfy local requests first. Naimi-Tréhel's algorithm was modified such that a request for entering a critical section (CS) follows the $LAST$ tree until it reaches the last node of the cluster which has currently requested the critical section, denoted the *local_root*. Hence, if the $NEXT$ of the *local_root* exists, it points to a site of a remote cluster. In this case, if the *local_root* receives a request from a node of its own cluster, and the number of preemptions is below the threshold, a local preemption of the token is performed and the *local_root*'s $NEXT$ is set to the local requester. The requesting node becomes the new *local_root*. In [Loallemi et al., 2006], Moallemi et al. propose to apply Bertier et al. approach to the fault tolerant Naimi-Tréhel [Sopena et al., 2005] in order to provide a fault tolerant hierarchical token-based mutual exclusion algorithm. They consider that at least one node of the cluster where the token is present does not crash. Broadcast was added between clusters for token and requests transmissions. No performance evaluation experiments were conducted.

In Mueller [Mueller, 1998], the author presents an extension to Naimi-Tréhel's algorithm where he adds the concept of priority in it. Local queues are introduced in each node. They form a virtual global queue ordered by priority and FIFO order is used when priorities are equal. A request is associated with a priority and the algorithm first satisfies the requests with higher priority. The basic idea is to accumulate priority information on intermediate nodes during request forwards. Thus, the $NEXT$ queue is replaced by a set of local queues. When a node issues a request with same priority, this request propagates along the $LAST$ tree until it reaches a node whose pending request is equal or greater than the former. The request is then locally queued at this node. When a node grants the token to another one it appends its local queue at the token. The receiving node then merges its token with the received one. In the author's approach, clustering is not considered. However, if priorities were assigned to requests based on their cluster locality, requests of the same cluster could be satisfied in sequence, preempting those of other clusters.

We should point out that the approach of changing the priority of pending requests could be applied to any other token-based algorithm. However, if such an approach allows to reduce the delay for a process to obtain the critical section, it does not reduce the complexity in terms of number of messages, which can become a drawback for large scale Grid systems.

2.3.2   Composition-based approach

Several authors [Chang et al., 1990a], [Housni and Tréhel, 2001], [Bertier et al., 2006], [Erciyes, 2004], [Madhuram and Kumar, 1994], [Omara and Nabil, 2002] propose to compose mutual exclusion algorithms: one at inter cluster/group level and a second at intra cluster/group level in order to minimize both message traffic and the time delay to enter the CS. In the majority of these works, the algorithm between clusters (groups) is different from the algorithm inside a group.

Chang et al. [Chang et al., 1990a] present in their article a hybrid approach which applies diffusion-based algorithms at both levels: Singhal's algorithm [Singhal, 1992] locally and Maekawa's algorithm [Maekawa, 1985] between groups. The former uses a dynamic information structure while the latter is based on a voting approach. The authors argue that since no algorithm can minimize both message traffic and time delay at the same time, such a combination is ideal since message complexity of Maekawa algorithm is low and Singhal's algortihm presents shorter delay time in successive executions of CS when compared to the former. Simulation studies show that when compared to flat Maekawa's algorithm, the proposed hybrid algorithm significantly reduces message traffic and time delay, specially if the system presents cluster locality of requests. Similarly, Omara et al.'s solution [Omara and Nabil, 2002] is a hybrid of Maekawa's algorithm at inter cluster level and Singhal's modified algorithm, which provides more fairness than the latter, at intra cluster level.

In Housni et al. [Housni and Tréhel, 2001], sites with the same priority are gathered within the same group. The algorithm between groups is different from the algorithm inside a group. They consider that routers ensure the interface between the algorithms of the two levels. Raymond's tree-based token algorithm [Raymond, 1989] is used inside a group, while Ricart-Agrawala [Ricart and Agrawala, 1981] diffusion-based algorithm is used between groups. If a router receives a permission from all the other routers, it generates the token in its group and gives it to the root node of its group. When all the pending requests of the group are satisfied, the token is restored to the router and destroyed. Routers can have different priorities. Hence, nodes of a given group, whose router has generated the token, can be preempted if its router receives a request from another one with higher priority. The former then interrupts the root of its group for requesting it the token. As soon as it gets it, the router sends its permission to the priority router.

Erciyes [Erciyes, 2004] proposes an architecture that consists of a logical ring of clusters, i.e., each node on the ring represents a cluster of nodes. Each of this node is a coordinator node, which performs the required critical section requests and releases on behalf of the nodes of the cluster it represents. However, inside the cluster, a central-based mutual exclusion algorithm is used, i.e., a node requests to its coordinator to enter the CS and will enter it upon reception of a reply message from the coordinator. When exiting the CS, the node sends a release message to the coordinator. The author proposes two algorithms for the inter-cluster level: the permission-based Ricart-Agrawala [Ricart and Agrawala, 1981] algorithm and Lan's token-based algorithm [Lann, 1978]. However, in both cases, messages are exchanged based on the inter cluster logical ring structure. Theoretical studies of the composed algorithms compared to their flat original algorithms show: (1) a gain of an order of magnitude of improvement in terms of message complexity but at the expense of larger obtaining time; (2) a reduced number of messages to enter a CS when the number of clusters increases. However, with a large number of clusters and number of messages (low level parallel application) coordinators become a bottleneck and token obtaining time increases.

Similarly to Erciyes [Erciyes, 2004], Madhuram et al. present in [Madhuram and Kumar, 1994] a two level algorithm where the centralized-based mutual exclusion algorithm is used at intra cluster level. However, coordinators at inter cluster level are not organized in a logical ring, but every coordinator can communicate with the other. The authors argue that the centralized algorithm is a good choice because of its low message complexity: 3 messages per CS invocation. All message requests are timestamped. The coordinator queues up the requests it receives from processes of its cluster in a queue in timestamped order (*InternalQ*). A second queue is used to store requests from the other coordinators (*GlobalQ*). A coordinator acts on behalf of all sites of its group. Upon receiving a request, the coordinator will broadcast a request provided it did not already do it (i.e., the request is registered in the *GlobalQ*). Using the *InternalQ*, the authors also propose a local preemption mechanism in order to satisfy local requests (imposing an upper limit to avoid starvation) before giving permission to a remote coordinator. They argue that in case of high load (low parallel application) such preemption mechanism increases performance of the application. The hierarchical algorithm presented in the paper adapts Ricart-Agrawala at inter-cluster level, but the authors advocate that any timestamp-based algorithms could also be used as well.

In [Bertier et al., 2006], Bertier et al. propose a composition approach which consists of Naimi-Tréhel algorithm at both level. The authors consider that messages exchanged by nodes of different clusters always pass along special nodes, called proxys, that behave like routers. There is one proxy per cluster. Therefore, every message that is sent from a node to a remote node is routed to the proxy of the cluster's sender which forwards the message to receiver's proxy node. The latter then sends the message to the receiver node. The sender's and receiver's proxies can then gather information about token transfers and requests at cluster level, taking decisions based on such information. Therefore, some inter-cluster messages can be managed at the proxies' level, without being necessary to be redirected to the other nodes inside clusters. Furthermore, by changing the $NEXT$ queue, intra-cluster CS requests are satisfied before remote ones. To avoid starvation, a threshold value limits the maximum number of CS that can be successively executed by nodes of the same cluster. Experiments were performed on a dedicated cluster where a Grid environment with multilevel network latencies was emulated by injection of network delays. The authors show that due to the proxy and cluster locality exploitation, both the time delay to obtain the token and the number of inter-cluster messages were reduced on their approach when compared to flat Naimi-Tréhel's algorithm, specially for low parallel applications.

All the above mentioned works hierarchically compose mutual exclusion algorithms but the choice for the composed algorithms can not change. One could argue that one of these fixed compositions could be more suitable for some type of application. However, no comparative evaluation performance study exists in the literature for such a conclusion. Another point is that the concept of group that the majority of these works exploit is logical, i.e., groups are not necessarily mapped to the physical topology and do not consider latency heterogeneity. Hence, whenever some performance results are presented in their respective articles, the experiments were conducted on top of simulators or emulated platform and not on top of a Grid.

2.4 Composition approach to mutual exclusion algorithms

Many of the composed algorithms presented in the previous section do not consider differences in communication latency as the main reason for grouping machines. Hence, the majority of them do not take into account the physical topology of the Grid and latency heterogeneity.

We have proposed a more generic approach which allows the choice of the good combination of algorithms according to the application's behavior by comparing different mutual exclusion algorithm compositions on top of Grid.

Similarly to a classical mutual exclusion algorithm, a mutual exclusion synchronization protocol on top of Grid should offer two operations: *CS_Request()*, which allows a process to request exclusive access to a shared resource, and *CS_Release()* called by the same process when it wants to release the resource. However, in order to be effective, such a synchronization protocol must tolerate heterogeneous network latencies. Our approach does it by using a hierarchy of mutual exclusion algorithms: a per cluster token-based mutual exclusion algorithm that controls critical section requests for processes within the same cluster and a second algorithm that controls *inter-cluster* requests for the token. The former is called the *intra* algorithm while the latter is called the *inter* algorithm and their executions are clearly separated. Furthermore, an *intra* algorithm instance of a cluster runs independently from a second *intra* algorithm instance. Thus, a process obtains access to the shared resource and later releases it by calling the above two operations which belong to the *intra* algorithm instance of its cluster. Another important advantage of our approach is that the chosen algorithms of both layers do not need to be modified. Hence, it is very simple to offer different implementations of a mutual exclusion service by just assembling multiple mutual exclusion algorithms.

Without loss of generality, we will consider that just one parallel or distributed application runs on top of the Grid at a given time. This application is composed of a set of processes and there is one process per node/machine denoted an *application* process.

When an *application* process wants to access the shared resource, it calls the operation *CS_Request()* of the *intra* algorithm. Upon getting the *intra* token, the process executes the critical section. After executing it, the process calls the operation *CS_Release()* of the same *intra* algorithm to release it. However, since one *intra* algorithm instance runs on each cluster, several processes could simultaneously access the critical section which would violate the safety property. In order to overcome this problem, we have introduced a special node within each cluster, called the *coordinator*, which ensures the safety property at a Grid wide level. The *inter* algorithm runs on top of the *coordinators* and allows a *coordinator* to request access to the shared resource on behalf of an *application* node within its respective cluster. *Coordinators* are in fact hybrid processes which participate in both the *inter* algorithm with the other *coordinators* and the *intra* algorithm with their cluster's *application* processes. Nevertheless, even if the *intra* algorithm sees a *coordinator* as an *application* process, the *coordinator* neither takes part in the application's execution nor requests access to the shared resource for itself.

The *coordinator* also uses the *CS_Request()* and the *CS_Release()* operations offered by the *inter* algorithm. However, a *coordinator* being in critical section for the *inter* algorithm instance means that one of the *application* processes of its cluster can access the resource. It is considered to be in the critical section by the other *coordinators*.

Each *intra* algorithm instance controls an *intra* token while the *inter* algorithm instances control an *inter* token. For every shared resource, there is one *intra* token per cluster but a single *inter* token for the whole system of which only the *coordinators* are aware. Therefore, holding the *intra* token of its cluster is sufficient and necessary for an *application* process to enter the CS since the local *intra* algorithm instance ensures that no other local *application* node of the cluster has the *intra* token. Moreover, considering the hierarchical composition of algorithms, our solution must also guarantee that no other *application* process of the other clusters is also in the critical section by holding the *intra* token of their own algorithm. In other words, at any time only one cluster has the right of allowing one of its *application* processes to execute the CS. This can be ensured by the possession of the *inter* token by just one of the *coordinators*.

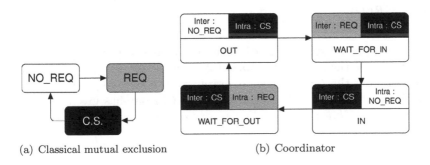

(a) Classical mutual exclusion   (b) Coordinator

FIGURE 2.4: Mutual exclusion automatas.

2.4.1   Coordinator processes

The guiding principle of our approach is represented by the automata of Figure 2.4.(b), which describes the behavior of a coordinator process. In a classical mutual exclusion algorithm, a process can be in one of the three following states: requesting the critical section (REQ), not requesting it (NO_REQ), or in the critical section (CS), as shown in Figure 2.4.(a). A *coordinator* process can also find itself in one of these three states, likewise the classical mutual exclusion, but with regard to both algorithms. Therefore, in the automata of Figure 2.4.(b), *Intra* and *Inter* refer to the *coordinator* state related to the *intra* algorithm instance and *inter* algorithm instance respectively. Moreover, a *coordinator* has additional states with respect to the *global state* of the composition, which can be one of the following: OUT, IN, $WAIT\_FOR\_OUT$,

WAIT_FOR_IN.

If the coordinator is in the *OUT* state, no local *application* process of its cluster has requested the CS. Thus, it holds the *intra* token (*Intra = CS*) but it does not hold the *inter* token (*Inter = NO_REQ*). When an *application* process wants to enter the critical section, it sends a request to the processes of its cluster by calling the *CS_Request()* operation of the *intra* algorithm. The *coordinator* of the cluster, which is the current holder of the *intra* token, will also receive such a request. Upon receiving it, the *coordinator* holder changes its global state to *WAIT_FOR_IN*. However, a *coordinator* can only grant the *intra* token to a requesting *application* process of its cluster if it holds the *inter* token too. To this end, it calls the *CS_Request()* operation of the *inter* algorithm in order to request the *inter* token. Therefore, when the *coordinator* finds itself in the *WAIT_FOR_IN* global state, there are one or more pending *intra* algorithm requests, and the *coordinator* still holds the local *intra* token (*Intra = CS*) but it is waiting for the *inter* token delivery (*Inter = REQ*).

The coordinator state changes to the *IN* global state when it receives the *inter* token. It then grants the *intra* token to the requesting *application* process by calling the *CS_Release()* operation of the *intra* algorithm. Thus, the coordinator holds the *inter* token (*Inter = CS*) but has given the *intra* algorithm token (*Intra = NO_REQ*) to one of the *application* processes of its cluster.

A *coordinator* which holds the *inter* token must also treat *inter* token requests received from the other coordinators. However, it can only grant the *inter* token if it also holds its local *intra* token. Holding this token ensures that there is no *application* process within its cluster in the critical section. When another *coordinator* requests the *inter* token, but the current holder of it does not keep both tokens, the latter sends a request to its *intra* algorithm asking for the *intra* token. Its global state triggers to *WAIT_FOR_OUT* coordinator, i.e., the *coordinator* still holds the *inter* token (*Inter = CS*) but it is waiting for the *intra* token (*Intra = REQ*) in order to be able to satisfy the *inter* algorithm pending requests. Upon obtaining the *intra* token, the *coordinator* can grant the *inter* token to the requesting *coordinator* by calling the *CS_Release()* operation of the *inter* algorithm. It returns to the *OUT* state where it holds the *intra* token (*Intra = CS*) but not the *inter* token (*Inter = NO_REQ*).

It is worth remarking that only one coordinator can be either in the *IN* or in the *WAIT_FOR_OUT* global state at any given time. All the other nodes are either in the *OUT* or in the *WAIT_FOR_IN* global state.

2.5   Composition properties and its natural effects

In order to show how much our composition approach reduces both the number of messages exchanged between application processes and the delay for a process to get access to the critical section, we are going to discuss an example of an application execution that uses our mutual exclusion composition protocol. Figure 2.5 shows such an example. It consists of a Grid composed of four three-node clusters. Each cluster i has two application nodes A_i and B_i and one coordinator node C_i. For the sake of simplicity, we consider that a token-based mutual exclusion algorithm that broadcast requests, such as Suzuki-Kasami algorithm [Suzuki and Kasami, 1985] (see Section 2.2), runs on both the *intra* and *inter* levels of the composition. All the effects described below then correspond to the gain in terms of number of messages and/or access waiting time of the composition protocol when compared to the original broadcast request mutual exclusion algorithm, i.e., if the latter, called the *flat* algorithm, was executed by all processes without a composition approach.

As we can observe in Figure 2.5(a), at the beginning of the execution, node A_1 keeps the *intra* token of cluster 1 and the coordinator C_1 keeps the *inter* token. *Intra* tokens of clusters 2, 3, and 4 are respectively held by coordinators C_2, C_3, and C_4.

2.5.1   Filtering and aggregation

Suppose that node B_3 wants to execute the critical section. It then broadcasts its request at *intra* level to all the nodes of its own cluster including the local coordinator C_3 (Figure 2.5(b)). Upon receiving the request, C_3 changes its state to $WAIT\_FOR\_IN$, forwards the request to all other coordinators and waits for the *inter* token. Only the coordinator C_1 will forward the request to the *application* nodes of its cluster by calling the $CS\_Request$ () operation of its *intra* algorithm. The other coordinators C_2 and C_4 will do nothing since they are in the OUT state and none of the *application nodes* of their clusters are in the critical section. The request was not broadcast in these two clusters and thus the number of *intra* cluster sent messages is reduced when compared to the *flat* algorithm. We denote such *intra* cluster message reduction the **natural filtering effect** of the composition.

A second effect of the composition is the **natural aggregation of the** *intra* **requests**. Before node B_3 gets the *intra* token, suppose that node B_2 calls the $CS\_Request()$ operation (Figure 2.5(c)). Hence, a request message is broadcast to all nodes of its cluster. When the coordinator C_2 receives the request, it forwards it to the other nodes of its cluster. However, since the coordinator C_1 is already in the $WAIT\_FOR\_OUT$ state due to B_3 request, it will not broadcast the request inside its cluster. Moreover, the request will not be broadcast inside those clusters where there is no node executing the

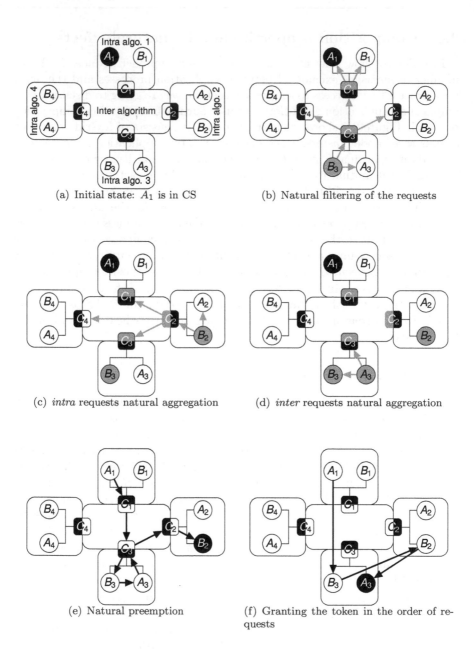

(a) Initial state: A_1 is in CS

(b) Natural filtering of the requests

(c) *intra* requests natural aggregation

(d) *inter* requests natural aggregation

(e) Natural preemption

(f) Granting the token in the order of requests

FIGURE 2.5: Example of execution.

critical, i.e., clusters C_3 and C_4.

Finally, suppose that a second *application* node of cluster 3, A_3, also decides to ask for the critical section. We observe in Figure 2.5(d) that the request is not broadcast at the *inter* level of the composition. This happens because the coordinator C_3 is in the $WAIT\_FOR\_IN$ state which means that it has already sent a previous *inter* request which has not been satisfied yet. Therefore, the composition approach naturally aggregates requests inside the same cluster which reduces the number of *inter* cluster messages. This effect is named the **natural aggregation of the** *inter* **requests**.

2.5.2   Preemption and structural effects

If we consider the previous execution, there are three concurrent pending requests which were respectively issued by B_3, B_2, and A_3. In the original *flat* algorithm, the requests would be satisfied in the order, i.e., first B_3's request, then B_2's request, and finally A_3's request. However, such an order forces the token to cross clusters (Figure 2.5(f)): from cluster 1 to cluster 3 when A_1 grants the token to B_3, from cluster 3 to cluster 2 when B_3 grants the token to B_2, and back to cluster 3 when B_2 grants the token to A_3. The round trip time for the token's travel between cluster 2 and 3 considerably increases the time for a process to get access to the critical section. Contrarily to the delay of a token request which can be overlapped by the duration of a critical section execution, no process can execute a critical section during the token transfer. Hence, reducing the delay of a token transfer has a direct impact on the overall performance of the algorithm since the time during which a node waits for the token will decrease as well.

Our composition protocol behaves differently (see Figure 2.5(e)). The hierarchy of algorithms naturally reorder the requests, giving priority to local requests over remote ones. When node A_1 ends its critical section, it calls *CS_Release()* and grants the *intra* token to C_1 which in its turn sends the *inter* token to C_3 by calling the *CS_Release()* of the *inter* algorithm. The coordinator C_3 then grants the *intra* token of cluster 3 to node B_3 (by calling *CS_Release()* of the *intra* algorithm) and changes its state from $WAIT\_FOR\_IN$ to IN. At the same time, thanks to the coordinator C_2, C_3 knows that there is a pending request from B_2, as explained above. Therefore, by calling the *CS_Request()* operation of the *intra* algorithm, C_3 broadcasts a request to all nodes of its cluster 3 in order to get the *intra* token. C_3 then changes its states to $WAIT\_FOR\_OUT$ state.

Notice that when node B_3 ends its critical section, there are two pending requests: one from A_3 and another from C_3. By keeping the order of reception of requests, B_3 grants the *intra* token to A_3 since it received A_3's requests before C_3's request. This means that without changing the original *intra* algorithm, the request of A_3 will be satisfied before the request of B_2 thus avoiding the aforementioned round trip travel time of the *inter* token between clusters 2 and 3. At the end of the critical section, A_3 grants the *intra* token to

C_3, since requests are satisfied in order. Having the *intra* token, C_3 grants the *inter* token to C_2 by calling the *CS_Release()* operation of the *inter* algorithm. Then, the nodes of cluster 2 can execute their critical section, i.e., B_2 in this example.

The reordering of the critical section accesses is due to the composition approach which prioritizes all the local requests of a cluster as soon as the latter gets the *inter* token. This mechanism optimizes the requests scheduling and can be seen as a "natural" scheduler because local requests preempt the critical section accesses over remote *application* node requests. Based on system taxonomy, we will refer to it as the "natural" preemption effect of the composition.

2.5.3   Natural effects of composition

Based on the previous example, we have discussed the effects induced by the mutual exclusion synchronization composition protocol. These effects are not due to the mechanisms added to coordinate at the different levels but they are a consequence of both the inherent hierarchical topology of the Grid and the use of coordinator processes. For these very reasons, we qualify them as "natural effects." These effects either reduce the number of sent messages or the time for a process to get to the CS.

Some of the above explained effects significantly reduce the mean number of sent messages which are required to get the critical section:

- **Natural filter effect**: requests are broadcast only in the last cluster which has executed the CS.

- **Natural aggregation of *inter* requests**: there is at most one broadcast in the *inter* algorithm for multiple concurrent requests issued from *application* nodes of the same cluster.

- **Natural aggregation of the *intra* requests**: inside the last cluster which has executed the CS, there is at most one broadcast of the *intra* algorithm of this cluster for all the requests forwarded by the other clusters.

The other effects directly reduce the mean CS access delay:

- **Natural preemption**: all pending requests of a cluster will be satisfied before the coordinator takes back the *intra* token and releases the *inter* token.

- **Structural effect**: In the *flat* algorithms, the delay for an *application* process to obtain the CS is proportional to the number of nodes of the whole system. On the other hand, the hierarchical structure of the Grid naturally reduces the number of application nodes in each instance of the algorithms. Therefore, the mentioned delay is reduced.

In conclusion, all theses effects improve the performances of the application over the *flat* algorithms. However, the overload of the composition should be taken into account, i.e., coordinators add a latency time and processor consumption to the system. The next section compares and shows the real gain of the hierarchical compositions over *flat* algorithms.

2.6   Performance evaluation

This section presents some performance evaluation results conducted on the French large-scale Grid'5000 [Cappello et al., 2006]. Some preliminary results have been presented in [Sopena et al., 2007] and [Sopena et al., 2008a].

Grid'5000 is a large-scale grid experimental testbed. It comprises 17 clusters located in 9 different cities all over France. Whichever the cluster, every node has a Bi-Opteron CPU and 2GB of RAM. Clusters are connected by dedicated 10Gb/s bandwidth links.

Our experiments used 9 of the 17 clusters, each one with 20 nodes, located in a different city. Figure 2.1 presents the average latency between the clusters.

Our performance tests aim at comparing the efficiency of some mutual exclusion algorithm compositions considering applications with different degrees of parallelism. The basic algorithms that we have chosen are **Martin**'s [Martin, 1985], **Naimi-Tréhel**'s [Naimi and Tréhel, 1996], and **Suzuki-Kasami**'s [Suzuki and Kasami, 1985] presented in Section 2.2.

2.6.1   Experiment parameters

The mutual exclusion algorithms as well as the coordinator are written in C using UDP sockets. An application process that runs on a single node executes 100 critical sections. Each of them lasts 10ms, which is the same order of magnitude as a data packet hop time between two clusters. Every experiment was executed 10 times and the presented results represent the average value.

An application behavior is characterized by:

- α: time taken by a node to execute the critical section;

- β: mean time interval between the release of the CS by a process and its next request.

- ρ: the ratio β/α, which expresses the frequency with which the critical section is requested.

We have developed several applications having **low, intermediate**, and **high** degrees of parallelism.

from \ to	Orsay	Grenoble	Lyon	Rennes	Lille	Nancy	Toulouse	Sophia	Bordeaux
Orsay	0.034	15.039	9.128	8.881	4.489	95.282	15.556	20.239	7.900
Grenoble	14.976	0.066	3.293	15.269	12.954	13.246	10.582	9.904	16.288
Lyon	9.136	3.309	0.026	12.672	10.377	10.634	7.956	7.289	10.078
Rennes	8.913	15.258	12.617	0.059	11.269	11.654	19.911	19.224	8.114
Lille	10.000	10.001	10.001	10.001	0.001	10.001	20.000	20.001	10.001
Nancy	5.657	13.279	10.623	11.679	9.228	0.032	98.398	20.001	12.827
Toulouse	15.547	10.586	7.934	19.888	19.102	17.886	0.043	14.540	3.131
Sophia	20.332	9.889	7.254	19.215	16.811	17.238	14.529	0.051	10.629
Bordeaux	7.925	16.338	10.043	8.129	10.845	12.795	3.150	10.640	0.045

Table 2.1: Grid'5000 RTT latencies (average *ms*).

Considering N as the total number of *application* processes (180 in our experiment), the three degrees of parallelism can be expressed respectively by:

- **Low Parallelism**: $\rho \leq N$: An application where the majority of *application* processes request the critical section. Thus, almost all *coordinators* wait for the *inter* token in the *inter* algorithm. In other words, almost all clusters have one or more *application* processes in the *requesting* state.

- **Intermediate parallelism**: $N < \rho \leq 3N$: A parallel application where some sites compete to get the CS. Only some *coordinators* are in the *requesting* state with respect to the *inter* algorithm on the whole Grid, i.e., just some clusters have one or more *application* that request the CS.

- **High Parallelism**: $3N \leq \rho$: A highly parallel application where concurrent requests to the CS are rare. The whole number of requesting *application* processes is small and usually distributed over the Grid. Hence, only one or a few clusters have one or more *application* processes in the *requesting* state with regard to the *inter* algorithm.

The performance of a mutual exclusion algorithm is usually measured by the number of messages exchanged per critical section and the delay for getting access to the shared resource, i.e., the time interval between the moment a node requests the CS and the moment it gets it. The latter, which we called the *obtaining time* in this paper, comprises the delay for transmitting a token request T_{req} plus the delay for granting the token T_{token}. However, if the time for waiting for the current pending requests T_{pendCS} is higher than T_{req}, the *obtaining time* is equal to T_{pendCS} plus T_{token}. Thus, the three metrics that we considered are: the **obtaining time**, i.e., the **number of sent messages**, and the **standard deviation** of the obtaining time.

For the sake of simplicity, we call the Naimi-Tréhel and Suzuki-Kasami algorithms respectively Naimi's and Suzuki's and for all figures of this section we have adopted the notation "*Intra algorithm-Inter Algorithm*" to denote a two level algorithm composition. For instance, "Naimi-Martin" denotes a composition where Naimi-Tréhel's algorithm is used as the *intra* algorithm of every cluster and Martin's algorithm as the *inter* algorithm.

2.6.2 Performance results: composition study

In this section we present evaluation performance results by composing the three algorithms described in Section 2.2 with different application behaviors.

The abscissae of the curves always represent the ρ parameter (degree of parallelism). Hence, when analyzing the curves the reader must keep in mind that when ρ increases, the number of processes that concurrently request the critical section decreases. As we observed that the *inter* algorithm has a much stronger influence in the overall performance than the *intra* algorithm, the experiments of Sections 2.6.2.1 and 2.6.2.2 have been performed by fixing the latter to Naimi's algorithm. Therefore, the variation of application processes *obtaining time* and number of *inter* cluster sent messages is only due to the *inter* algorithm. The latter comprises the number of messages for delivering *inter* token requests plus the number of messages for granting the *inter* token. The impact of the *intra* algorithm choice on the overall performance of our composition approach as well as the advantages of choosing Naimi's for the *intra* algorithm are explained in Section 2.6.2.4.

2.6.2.1 Obtaining time of application processes

We consider the following notations:

- T: average message delay for transmitting a message between two coordinators;

- T_{req}: average message delay for transmitting an *inter* token request message from a coordinator to the coordinator that will grant it the token.

- T_{token}: average message delay for granting the token between the current coordinator token holder and the requesting coordinator;

- T_{pendCS}: average delay for satisfying all the current pending *inter* token requests before satisfying the studied *inter* token request.

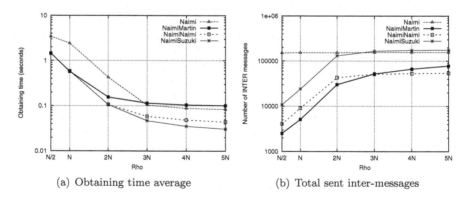

(a) Obtaining time average   (b) Total sent inter-messages

FIGURE 2.6: Composition evaluation.

In terms of *obtaining time*, a first remark is that for all curves the obtaining latency decreases with the decreasing of concurrency, i.e., the reduction of the waiting queue size. The clustering of *intra* token requests has also an advantageous impact on the *obtaining time* when compared to the original algorithm, as we can observe in Figure 2.6.(a). Such a benefit depends on ρ.

In highly parallel applications where there is almost no concurrency among accesses to the shared resource ($\rho \geq 3N$), the *obtaining time* of a coordinator comprises the request message delay T_{req} plus the token message delay T_{token}. However, in applications with high concurrency for accessing a shared resource, as in low parallel applications ($\rho \leq N$), a coordinator must wait for all the other pending CS requests to be satisfied before getting the token. This delay, which we called T_{pendCS}, is usually higher than the one for sending the request T_{req} and completely overlaps T_{req}. Therefore, the *obtaining time* of a coordinator consists of T_{pendCS} plus the token message delay T_{token}. This explains why the *obtaining time* tends to be higher when $\rho \leq N$, since in this case there are always many application processes in the *requesting* state, and quite short when $\rho \geq 3N$, since the number of waiting coordinators for the token is small. Such a behavior can be observed in Figure 2.6.(a).

Low parallel application: We did not observe any significant difference with respect to the average *obtaining time* of all three algorithms of Figure 2.6.(a) for $\rho \leq N$. As explained above, in this case, the *obtaining time* of a coordinator is equal to T_{pendCS} plus T_{token}. T_{pendCS} is the same for all three *inter* algorithms while T_{token} is reduced to T in the case of Naimi's (a send to the *next* node) and Suzuki's (a send to the first node of Q) algorithms. In Martin's algorithm, the current token holder grants the token to its predecessor in the ring. However, as this node has a very high probability of having requested the token too, the token granting delay also takes one message ($Ttoken = T$), as in the other two algorithms.

As concurrency among accesses to the shared resource is quite high in low parallel applications, the *obtaining time* does not vary much. Such a behavior will be explained in Section 2.6.2.3, where the standard deviation of the *obtaining time* is discussed.

Intermediate parallel application: A first remark is that Naimi-Naimi's *obtaining time* is comparable to Naimi-Suzuki's (see Figure 2.6.(a) for $N < \rho \leq 3N$) whereas Naimi-Martin's is slightly higher. This is explained by the fact that when using Martin's as the *inter* algorithm, there are some coordinators waiting for the *inter* token which implies that their T_{req} can still be covered up by their T_{pendCS}. Thus, similarly to low parallel applications, the main factor for the *obtaining time* is T_{token}. Suzuki's and Naimi's *inter* algorithm invariably need only one message, whose delay is T while Martin's needs more than one message in average. For Martin's, the smaller is the number of pending requests, the lower is the probability that a second coordinator has also requested the token and the higher is the probability that

T_{token} increases. Therefore, Martin's algorithm is not suitable as the *inter* algorithm for this type of application.

Highly parallel application: In the case of applications with high degree of parallelism, CS requests from *application* processes are quite sparse. As explained above, in such applications, the *obtaining* time of a coordinator comprises the requesting message delay T_{req} plus the *inter* token message delay T_{token}. As the application does not present much concurrency, T_{token} is equal to T to both Naimi's and Suzuki's algorithms while for Martin's it is equal to $N/2 * T$.

In terms of T_{req}, the most effective *inter* algorithm is Suzuki's, since a CS requesting is performed by a single message sent in parallel to each coordinator, taking just T. As Naimi's uses a tree to route requests, the average delay for a request travel is $log(N) * T$ between coordinator nodes. The less suitable algorithm is Martin's. Since the number of requesting coordinator tends to zero, a CS request tends to travel along the ring an average of $N/2$ successive hops, which implies a T_{req} of $N/2 * T$. Hence, the impact of T_{req} in the *obtaining* time of the three algorithms explains why Suzuki's presents the lowest *obtaining* time and Martin's the highest one as observed in Figure 2.6.(a) for $\rho \geq 3N$,

We can summarize our study about the *obtaining* time by the Table 2.2.

Composition \ Parallelism	Low	Intermediate	High
Naimi-Suzuki	$T_{pendCS} + T$	$T_{pendCS} + T$	$T + T$
Naimi-Martin	$T_{pendCS} + T$	$T_{pendCS} + K * T$	$N * T$
Naimi-Naimi	$T_{pendCS} + T$	$T_{pendCS} + T$	$log(N) * T + T$

Table 2.2: Average token obtaining time per composition.

2.6.2.2 Number of inter-cluster sent messages

In Figure 2.6.(b), we can see that, independently of ρ, the original Naimi-Tréhel always presents the same number of *inter* cluster sent messages ($\mathcal{O}(log(N))$). This constant behavior can be explained since the routing of both a CS request and a token granting message from a node does not depend on its location. A message is arbitrarily routed through nodes which are within the same cluster or belong to different clusters. On the other hand, when a compositional approach is used, *inter* cluster messages are managed by coordinators which gather token request messages from *application* processes into just one *inter* token request. Hence, the number of *inter* cluster sent messages decreases when compared to the original algorithm, as we can observe in the same figure for all three algorithm compositions. Nevertheless, when applying our composition approach, the number of *inter* cluster sent messages is not constant but increases with ρ.

When ρ is small, there is a lot of concurrent CS requests from *application* processes of the same cluster which result in a single *inter* token request by the coordinator of the cluster in question. In this particular case, we should emphasize the advantage of using the Naimi-Naimi's algorithm composition compared to the original one. But when concurrency for the CS decreases, the gathering of *intra* CS requests by a coordinator decreases as well which implies results in more *inter* cluster requests.

In the case of Suzuki's and Naimi's inter algorithms, the number of sent messages per *inter* token request of a coordinator consists of one message for the grant of the *inter* token and respectively N messages and $\mathcal{O}(log(N))$ messages for *inter* token request. Hence, in terms of number of *inter* cluster sent messages, Naimi's is more efficient than Suzuki's, which can be observed in the curves Naimi-Naimi and Naimi-Suzuki of Figure 2.6.(b). However, in the case of Martin's algorithm, that number depends on ρ. For low parallel applications ($\rho \leq N$), the probability of having all coordinators requesting the *inter* token at a given time is high. Therefore, the grant of the *inter* token takes just one message as well as a coordinator request since a second coordinator which is in a *requesting* state does not forward a request, as explained in 2.2.1. When the parallelism of the application increases, the number of inter-cluster sent messages per *inter* token request increases as well. This growth can be explained since the probability that some coordinator requests the *inter* token decreases. Thus, the number of hops of a request message increases proportionally, which generates more messages. In a highly parallel application, a token request in Martin's generates $N/2$ messages and the grant of the token generates $N/2$ messages. By comparing Naimi-Martin and Naimi-Naimi curves of Figure 2.6.(b), we can observe that for highly parallel applications ($\rho \geq 3N$), the number of *inter* cluster messages sent by Martin's is slightly higher than Naimi's.

The number of *inter* cluster messages can be summarized by the Table 2.3.

Parallelism Composition	Low	Intermediate	High
Naimi-Suzuki	$N+1$	$N+1$	$N+1$
Naimi-Martin	2	$2 < K < N$	$(N/2)+(N/2)$
Naimi-Naimi	$log(N)+1$	$log(N)+1$	$log(N)+1$

Table 2.3: Average number of inter cluster messages per composition.

2.6.2.3   Standard deviation

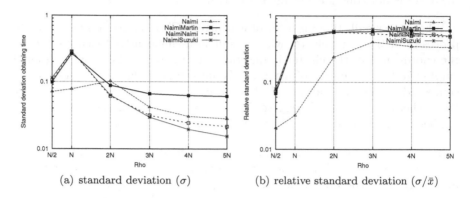

(a) standard deviation (σ) (b) relative standard deviation $(\sigma/\bar{x})$

FIGURE 2.7: Obtaining time standard deviation.

In order to analyze more precisely the variation of the *obtaining time*, its standard deviation σ has been measured, as shown in Figure 2.7.(a). A first remark when observing this figure is that σ is in fact quite significant for all ρ values compared to the average CS time. This is due to the communication heterogeneity of the Grid platform: inter cluster latencies are much higher than intra cluster ones and the former are not uniform with regard to two different clusters, as described in Figure 2.1.

To measure the importance of σ and to evaluate the side effects of the average *obtaining time* variations, we choose to study the relative deviation time $\sigma_r = (\sigma/\bar{x})$, which is the ratio of the standard deviation σ to the average *obtaining time* $\bar{x}$ – see Figure 2.7.(b). The original Naimi's algorithm relative deviation σ_r is always smaller than that of any composition of algorithms. This happens because in the case of Naimi's, the path covered by the token is independent of the actual token position. However, in our approach, a request can have one of the following two delays: a very short one when the token is already in the cluster of the requesting node, and a long one when the token is not in the same cluster.

All curves of the Figure 2.7.(b) have the same form: a significant growth for the lower values of ρ and then a stabilizing phase. This growth of σ_r can be explained by two phenomena: the overlapping of the requesting trip time (T_{req}) by the process time of the requesting queue and the sequential ordering due to the extreme number of requests (for $\rho = N/2$).

With respect to the difference between the compositions curves, we can note that they are equivalent for lower values of ρ. For the intermediate parallel degree $(N < \rho \leq 3N)$, Naimi-Martin's has the worst absolute standard deviation due to its logical ring structure while Naimi-Suzuki's and Naimi-Naimi's present a better absolute standard deviation. However, Naimi-Suzuki exhibits a better relative standard deviation. For $\rho > 3N$, Naimi-Suzuki has the smallest σ as shown in Figure 2.7.(a).

2.6.2.4   Intra algorithm choice

We have carried out several experiments aiming at choosing the best *intra algorithm* with respect to the behavior of the applications. In order not to load Figure 2.8, we just show the curves when the *inter* algorithm is fixed to Naimi's. Experiments with the other two algorithms have presented the same behavior.

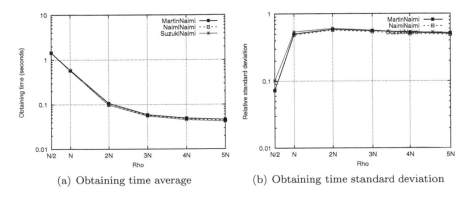

(a) Obtaining time average   (b) Obtaining time standard deviation

FIGURE 2.8: Intra algorithm.

In terms of the number of *intra* cluster messages, all algorithms have an acceptable local overhead. One could argue that since Suzuki's algorithm sends a much higher number of request messages per critical section than the other two algorithms, it might be not chosen as the *intra* algorithm. However, as nodes within a cluster are linked by a LAN, a multicast primitive could be used to diffuse the request which will significantly reduce the number of sent messages.

Concerning the *obtaining time* (Figure 2.8.(a)), all algorithms present almost the same curve, independently of ρ with a slight advantage for Suzuki-Naimi. Still, the latter has a weaker regularity (Figure 2.8.(b)) than Naimi-Naimi. This difference is due to the lack of fairness of Suzuki's algorithm when appending nodes to the token queue Q since it does not consider the arrival time of the requests.

Therefore, the regularity and performance of Naimi's algorithm justify choosing it as the intra algorithm in the experiments of the previous sections.

2.6.3   The impact of the grid architecture

The current performance evaluation aims at studying and comparing the influence of the Grid architecture in both the original Naimi-Tréhel mutual exclusion algorithm (*flat* algorithm) and in our composition approach using Naimi-Tréhel at both levels (*hierarchical* algorithm). To this end, the number of nodes of the Grid was set to 120 but the number of clusters varied: 2, 3, 4, 6, 8, 12, 20, 30, 40, 60, and 120. The experiments were conducted on a dedicated cluster of twenty-four Bi-Xeon 2.8 Ghz with 2GB of RAM machines where a Grid environment with 120 virtual nodes was emulated. There is one process per virtual node. For those configurations where the number of virtual clusters is greater than the number of available machines, nodes of the same virtual cluster run on the same machine. This approach prevents side effects of *intra* cluster communication.

Network latencies between clusters were emulated by using the flexible tool DUMMYNET [Rizzo, 1997] which allows injection of network delay, bandwidth limitation, and packet loss. Hence, for emulating several virtual clusters, every message exchanged between two virtual clusters goes through a dedicated machine, a P4 3Ghz machine, which runs a FreeBSD DUMMYNET. *Intra* cluster communication latency is 0.5ms while *inter* cluster latency is 20ms. Machines are connected by a 140 Gbits/s Ethernet switch.

In order to evaluate the *flat* algorithm as well as the *hierarchical* one, two metrics have been considered: (1) the *number of inter-cluster messages* and (2) the *obtaining time*.

Considering $N = 120$, for each experiment, we have measured the *obtaining time* (Figures 2.9(a), 2.9(b), and 2.9(c)) and the number of *inter* cluster messages (Figures 2.9(d), 2.9(e), and 2.9(f)) for both algorithms when the number of cluster ranges from 2 to 120. Figures 2.9(a) and 2.9(d) correspond to a low parallel degree application ($\rho = N/2$); Figures 2.9(b) and 2.9(e) correspond to an intermediate parallel degree application ($\rho = 2N$); Figures 2.9(c) and 2.9(f) correspond to a high parallel degree application ($\rho = 5N$).

2.6.3.1   Flat algorithm

We start by studying the impact of the number of clusters of the Grid on both the *obtaining time* and the number of *inter* cluster messages in the original *flat* Naimi-Tréhel algorithm. We can observe in Figure 2.9 that the curves related to this algorithm have a quite similar form. Independently of ρ, all curves present a hyperbolic form: a significant growth when the number of clusters varies from 2 to 12. This growth is then strongly reduced, becoming almost null, when the number of clusters is greater than 40.

In order to explain the form of such curves, we propose to theoretically study the frequency with which a *flat* mutual exclusion algorithm sends an *inter* cluster message, i.e., the probability $\mathcal{P}$ that the destination node of a message does not belong to the same cluster of the message's sender. To this end, we consider a Grid architecture composed of N nodes uniformly distributed over c clusters. Without loss of generality, we also suppose that a node can send a message to itself. This assumption models two successive accesses to the critical section by the same node. Then, we get the following probability $\mathcal{P}$:

$$\mathcal{P} = \frac{N - \frac{N}{c}}{N} = 1 - \frac{1}{c}.$$

This equation is totally in accordance with the form of the curves of Figures 2.9 for the *flat* algorithm. It also shows that such a probability does not depend on the number of nodes N whenever they are uniformly distributed over the Grid, i.e., it depends only on c. A last important conclusion from this equation is that the clustering effect due to the communication latency heterogeneity of a Grid has a negligible impact on the order of CS accesses. In other words, such a heterogeneity does not change the order of priority of the requests in such a way that request from closer nodes would be satisfied before distant ones.

In the above equation, any node can be chosen among N with the same probability, independently of the Grid topology. Furthermore, if theoretical curves were drawn from the equation, they would be similar to the ones of Figure 2.9. Thus, we can deduce that the assumption of equiprobability is reasonable and that the algorithm does not naturally adapt itself to the Grid topology.

Let's come back to the curves in order to study the impact of the number of clusters with respect to the application behavior. The results of Figures 2.9(a), 2.9(b), and 2.9(c) show that the degree of parallelism of an application has an impact on the *obtaining time*. Furthermore, the curves of Figures 2.9(d), 2.9(e), and 2.9(f) show that the parallelism degree of an application has no influence on the number of *inter* cluster messages even if we observe a small reduction of this number for low parallel applications.

2.6.3.2   Hierarchical algorithm

We are now going to study the impact of the Grid architecture on our hierarchical approach. The number of clusters has an influence on the *obtaining time* as well as in the number of *inter* clusters which increase with the number of clusters. However, if we exclude the configuration with one node per cluster where there is in fact no hierarchy of communication at all, our approach always presents a smaller *obtaining time* and number of *inter* cluster messages when compared to the *flat* algorithm. Notice that the benefit of using our composition approach is considerable even for a Grid composed of 60 two-node clusters.

Since the topology of the Grid has not the same impact on our composition approach as on the *flat* algorithm, it would be interesting to study the mean deviation between the *hierarchical* curves and the *flat* ones for both the *obtaining time* and the number of *inter* cluster messages. Thus, based on the curves of Figure 2.9, Figure 2.10 shows such mean deviations.

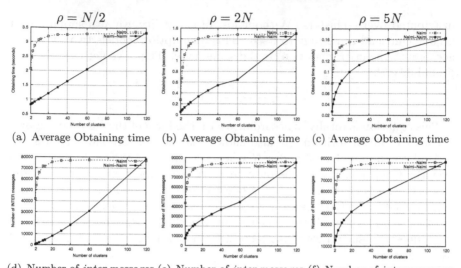

(a) Average Obtaining time (b) Average Obtaining time (c) Average Obtaining time

(d) Number of *inter* messages (e) Number of *inter* messages (f) Number of *inter* messages

FIGURE 2.9: Impact of the number of clusters.

In Figure 2.10, we can observe that the gain of our composition approach increases when the number of clusters ranges from 2 to 12. This is in accordance with the curves of Figures 2.9 where the *obtaining time* as well the number of *inter* cluster messages increase sharply for the original algorithm but smoothly for our composition approach. Such a different behavior explains why the maximum mean deviation between the two curves is reached with 12 clusters. Beyond this threshold value, the clustering effect neither has an influence on the *obtaining time* nor on the number of *inter* cluster messages since in our *hierarchical* approach the curves progressively increase while in the curves of the *flat* algorithm remain linear. Thus, the respective mean deviations inversely decrease until they become null for the configuration where each node represents a cluster (120 clusters).

We would like to theoretically evaluate the above threshold in a Grid composed of N nodes uniformly divided into c clusters. Hence, similarly to Section 2.6.3.1, we need to find the probability $\mathcal{P}$ that a node sends an *inter* cluster message in our own hierarchical approach on top of such a Grid. Without loss of generality, we consider the case where the cluster locality is maximum, i.e., every time a coordinator of a cluster gets the *inter* token, all the N/c nodes of this cluster execute a critical section which corresponds to a low parallel application. Thus, the probability $\mathcal{P}$ is equal to the probability of executing the last of the N/c critical section executions:

$$\mathcal{P} = \frac{1}{\frac{N}{c}} = \frac{c}{N}$$

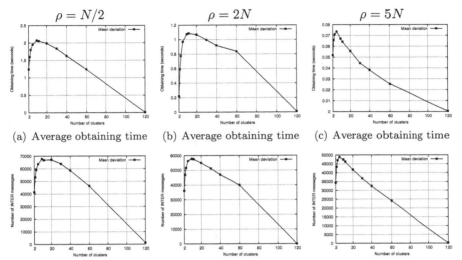

(a) Average obtaining time   (b) Average obtaining time   (c) Average obtaining time

(d) Number of *inter* messages (e) Number of *inter* messages (f) Number of *inter* messages

FIGURE 2.10: Mean deviation between the composition approach and the original algorithm.

Therefore, the mean deviation $E(c)$ between our composition approach and the *flat* algorithm in function of the number of clusters c is equal to:

$$E(c) = 1 - \frac{1}{c} - \frac{c}{N}$$

and according to the derivative of E, the mentioned threshold, $c_{threshold}$, is equal to:

$$E'(c) = \frac{1}{c^2} - \frac{1}{N} = 0 \Rightarrow c_{threshold} = \sqrt{N}$$

Such an equation shows that the maximum benefit when using our composition approach is reached for a Grid architecture composed of $\sqrt{N}$ nodes. This result can be verified by the curves of Figure 2.10 since $\sqrt{120} = 10.95$. Consequently, for $\rho = N/2$ and $\rho = 2N$, the maximum mean deviation is reached between 8 and 12 clusters. It is also worth noting that for low parallel applications ($\rho = 5N$), the Grid architecture corresponding to the highest benefit is equal to 6 clusters.

Finally, contrarily to the *flat* algorithm, the parallelism degree of an application has an influence on our hierarchical approach. Indeed, we can observe in the curves of Figure 2.9 that it becomes less effective with higher parallel applications when the number of clusters increases, i.e., it does not present a linear behavior anymore as it does with low parallel applications.

2.6.3.3 Heterogeneous grid architecture

In the previous section we have presented performance results related to homogenous Grid architectures, i.e., topologies where the clusters have the same number of nodes. However, in the case of heterogenous architecture, it is very difficult to conduct the same performance experiments due to the great number of possible topologies. Thus, we are going to present in this section just a theoretical study. To this end, we consider a Grid composed of n nodes and c clusters such that cluster i contains m_i nodes ($1 \leq i \leq c$). Notice that the probability that two nodes of different clusters communicate does not depend on the number of clusters c anymore, but it is a function of the distribution of the nodes among the c clusters. We denote M such a distribution and we denote $\mathcal{M}_c$ the set of such distributions. Thus, we have:

$$\mathcal{P}(M) = \sum_{i=1}^{c} \left(\frac{m_i}{n} \right) \left(\frac{n - m_i}{n} \right)$$

$$= \frac{1}{n^2} \sum_{i=1}^{c} (m_i n) - \frac{1}{n^2} \sum_{i=1}^{c} m_i^2$$

$$= 1 - \frac{\sum_{i=1}^{c} m_i^2}{(\sum_{i=1}^{c} m_i)^2}$$

The probability that a message travels along the WAN (inter-cluster network) is thus proportional to the ratio of the sum of squares to the square of the sum. Such a probability can then be raised by the Cauchy-Schwarz inequality ([Schwarz, 1888]):

$$\text{Cauchy-Schwarz inequality} \Rightarrow \left(\sum_{i=1}^{c} m_i \times 1 \right)^2 \leq \left(\sum_{i=1}^{c} m_i^2 \right) \left(\sum_{i=1}^{c} 1^2 \right)$$

$$\Rightarrow \left(\sum_{i=1}^{c} m_i \right)^2 \leq c \sum_{i=1}^{c} m_i^2$$

$$\Rightarrow \frac{1}{c} \leq \frac{\sum_{i=1}^{c} m_i^2}{(\sum_{i=1}^{c} m_i)^2}$$

$$\Rightarrow \forall c, \ \forall M \in \mathcal{M}_c, \ \mathcal{P}(M) \leq 1 - \frac{1}{c}$$

Hence, the probability of sending a message over the WAN for the set of configurations $\mathcal{M}_c$ which contains c clusters can also be raised by the constant $1 - 1/c$. Such a value holds when a homogenous configuration $H_c \in \mathcal{M}_c$ is considered (see Section 2.6.3.1). We have thus the following threshold:

$$\forall c, \ \forall M \in \mathcal{M}_c, \ \mathcal{P}(M) \leq \mathcal{P}\left(H_c \right)$$

We are going now to study how the probability $\mathcal{P}(M)$ varies according to different heterogeneous Grid topologies for a given number of clusters. The goal of this study is to analyze how fast the cost of using a Grid converges to its maximum (homogenous clustering architecture).

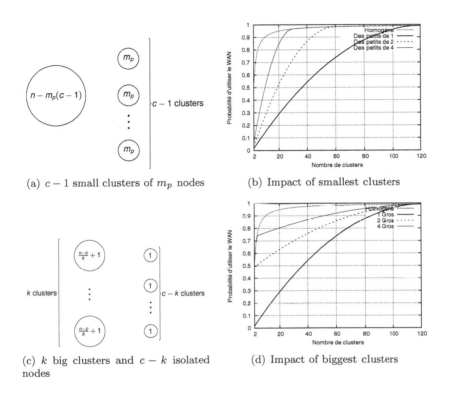

(a) $c-1$ small clusters of m_p nodes   (b) Impact of smallest clusters

(c) k big clusters and $c-k$ isolated nodes   (d) Impact of biggest clusters

FIGURE 2.11: Probability of using the WAN for different grid topologies.

Firstly, we consider a Grid composed of a single big cluster of size m_b, and $c-1$ small ones of size is m_s. (Figure 2.11(a)). Since the number of nodes does not change, the size m_b of the big cluster depends on c and m_s, i.e., it is equal to the number of machines that have been not distributed among the $c-1$ small clusters: $m_b = n - m_s(c-1)$. Figure 2.11(b) shows the probability of sending a message over the WAN for four configuration: $m_s = 1$, $m_s = 2$, $m_s = 4$, and $m_s = n/c$ (homogenous configuration). The curves represent the impact of the minimum size of the clusters on the performance of the algorithms. We could note that when there are $n-1$ clusters composed of one node, the probability of using the WAN increases almost linear when the number of clusters increases as well. On the other hand, if the size of the small clusters slightly increases, such a probability increases very fast.

We are going to now study the probability of sending a message over the WAN on a topology composed of k big clusters of the same size m_b. The other ones are small clusters of size m_s composed of just one node, i.e., $m_s = 1$ (Figure 2.11(c)). It is worth remarking that the size m_b of the k big clusters reduces when k increases: $m_b = (n - c)/k + 1$. Figure 2.11(d) shows the probability of sending a message over the WAN for four configurations: $k = 1$, $k = 2$, $k = 4$, and $k = c$ (homogenous configuration). Similarly to the previous results, the curves prove that by increasing the number of clusters, the same results obtained for homogeneous configurations hold very fast. For instance, having four big clusters ($k = 4$) among 20 clusters, the probability of sending a message over the WAN is 80%.

Based on both theoretical studies, we can conclude that if we exclude the case of a Grid composed of a single big cluster and $n - 1$ small clusters of one node, the probability of sending a message over the WAN on a heterogeneous Grid topology tends to be the same as the one on a homogeneous Grid topology.

2.7   Concluding remarks

We have proposed a new approach for composing mutual exclusion algorithms in order to offer mutual exclusion service for Grid environments where application processes are spread over several clusters interconnected by long distance links. Such a composition is totally transparent to the application and any classical token-based algorithm can be chosen as both inter and intra algorithms. Our two-level approach is scalable and can be easily extended to multiple levels of algorithm hierarchy which render it extremely suitable for large-scale systems.

Performance evaluation results from experiments conducted on both the real Frenchwide Grid'5000 and an emulation platform show that the degree of parallelism of an application has an impact on the choice of the *inter* algorithm. Such a choice depends on the logical topology that the algorithm takes into account for forwarding the token request. To this end, Martin's, Naimi-Tréhel's, and Suzuki-Kasami's algorithms which respectively consider a ring, a tree, and a complete graph topology were used as the *inter* algorithm in our tests. When the system is stressed (the rate of CS request is high and there are requests in all clusters), a ring topology is the most effective; when the CS rate is lower (i.e., the application exhibits a higher degree of parallelism) both the tree and the complete graph configurations are more efficient since they reduce the number of hops of CS request messages. Such results prove that our approach provides a framework for easily choosing the best two algorithms combination for composing mutual exclusion services.

2.8 References

[Baarir et al., 2008] Baarir, S., Sopena, J., and Legond, F. (2008). On the formal verification of a generic hierarchical mutual exclusion algorithm. In *Proceedings of the 28th International Conference on Formal Techniques for Networked and Distributed Systems (FORTE'08)*, Lecture Notes in Computer Sciences, Tokyo, Japan. Springer-Verlag.

[Bertier et al., 2004] Bertier, M., Arantes, L., and Sens, P. (2004). Hierarchical token-based mutual exclusion algorithms. In *Proceedings of the 4th IEEE/ACM International Symposium on Cluster Computing and the Grid*. IEEE Computer Society.

[Bertier et al., 2006] Bertier, M., Arantes, L., and Sens, P. (2006). Distributed mutual exclusion algorithms for grid applications: a hierarchical approach. *Journal of Parallel and Distributed Computing*, 66:128–144.

[Cappcllo et al., 2006] Cappello, F., Desprez, F., Dayde, M., Jeannot, E., Jegou, Y., Lanteri, S., Melab, N., Namyst, R., Primet, P. V.-B., Richard, O., Caron, E., Leduc, J., and Mornet, G. (2006). Grid5000: a nation wide experimental grid testbed. *International Journal on High Performance Computing Applications*, 20(4):481–494.

[Chang et al., 1990a] Chang, I., Singhal, M., and Liu, M. (1990a). A hybrid approach to mutual exclusion for distributed system. In *Proceedings of the IEEE International Computer Software and Applications Conference*, pages 289–294. IEEE Computer Society.

[Chang et al., 1990b] Chang, I., Singhal, M., and Liu, M. (1990b). An improved $o(log(n))$ mutual exclusion algorithm. In *Proceedings of the 1990 International Conference on Parallel Processing*, pages 295–302.

[Erciyes, 2004] Erciyes, K. (2004). Distributed mutual exclusion algorithms on a ring of clusters. In *Proceedings of the International Conference on Computational Science and its Applications*, volume 3045 of *Lecture Notes in Computer Sciences*, pages 518–527.

[Housni, 2002] Housni, A. (2002). *Introduction de la priorité dans les algorithmes d'exclusion mutuelle répartis*. PhD thesis, Université de Franche-Comté.

[Housni and Tréhel, 2001] Housni, A. and Tréhel, M. (2001). Distributed mutual exclusion by groups based on token and permission. In *International Conference on Computational Science and its Applications*, pages 26–29.

[Lamport, 1978] Lamport, L. (1978). Time, clocks, and the ordering of events in a distributed system. *Communications of ACM*, 21(7):558–565.

[Lann, 1978] Lann, G. L. (1978). Algorithms for distributed data-sharing systems which use tickets. In *Proceedings of the 3rd Berkeley Workshop on Distributed Data Management and Computer Networks*, pages 259–272.

[Loallemi et al., 2006] Loallemi, M., Mansouri, Y., Rasoulifard, A., and Naghibzadeh, M. (2006). Fault-tolerant hierarchical token-based mutual exclusion algorithm. In *Proceedings of the International Symposium in Communications and Information Technologies*, pages 171–176.

[Madhuram and Kumar, 1994] Madhuram and Kumar (1994). A hybrid approach for mutual exclusion in distributed computing systems. In *Proceedings of the IEEE Symposium on Parallel and Distributed Processing*. IEEE Computer Society.

[Maekawa, 1985] Maekawa, M. (1985). A square root n algorithm for mutual exclusion in decentralized systems. *ACM Transactions on Computer Systems*, 3(2):145–159.

[Martin, 1985] Martin, A. J. (1985). Distributed mutual exclusion on a ring of processes. *Science of Computer Programming*, 5(3):265–276.

[Mueller, 1998] Mueller, F. (1998). Prioritized token-based mutual exclusion for distributed systems. In *Proceedings of the International Parallel Processing Symposium*, pages 791–795.

[Naimi and Tréhel, 1996] Naimi, M. and Tréhel, M. (1996). A $log(n)$ distributed mutual exclusion algorithm based on the path reversal. *Journal of Parallel and Distributed Computing*, 34:1–13.

[Omara and Nabil, 2002] Omara, F. and Nabil, M. (2002). A new hybrid algorithm for the mutual exclusion problem in the distributed systems. *International Journal of Intelligent Computing and Information Sciences*, 2(2):94–105.

[Raymond, 1989] Raymond, K. (1989). A tree-based algorithm for distributed mutual exclusion. *ACM Transactions on Computer Systems*, 7(1):61–77.

[Ricart and Agrawala, 1983] Ricart, G. and Agrawala, A. (1983). Author response to 'on mutual exclusion in computer networks' by Carvalho and Roucairol. In *Communications of ACM*, volume 26/2, pages 147–148.

[Ricart and Agrawala, 1981] Ricart, G. and Agrawala, A. K. (1981). An optimal algorithm for mutual exclusion in computer networks. *Communications of ACM*, 24(1):9–17.

[Rizzo, 1997] Rizzo, L. (1997). Dummynet: a simple approach to the evaluation of network protocols. *ACM Computer Communication Review*, 27(1):31–41.

[Schwarz, 1888] Schwarz, H. A. (1888). Ueber ein flachen kleinsten flachen-inhalts betreffendes problem der variationsrechnung. *Acta Societatis scientiarum Fennicae*, XV:318.

[Singhal, 1992] Singhal, M. (1992). A dynamic information structure for mutual exclusion algorithm for distributed systems. *IEEE Transactions on Parallel and Distributed Systems*, 3(1):121–125.

[Sopena et al., 2005] Sopena, J., Arantes, L., Bertier, M., and Sens, P. (2005). A fault-tolerant token-based mutual exclusion algorithm using a dynamic tree. In *Proceedings of the Euro-Par 2005 Conference on Processing*, volume 3648/2005, pages 654–663, Heidelberg, Deutschland. Springer-Verlag.

[Sopena et al., 2008a] Sopena, J., Arantes, L., Legond, F., and Sens, P. (2008a). The impact of clustering on token-based mutual exclusion algorithms. In *Proceedings of the Euro-Par 2008 Conference on Processing*, volume 3648/2005, Heidelberg, Deutschland. Springer-Verlag.

[Sopena et al., 2009a] Sopena, J., Arantes, L., Legond, F., and Sens, P. (2009a). Building effective mutual exclusion services for grids. *Journal of Supercomputing*.

[Sopena et al., 2006a] Sopena, J., Arantes, L., and Sens, P. (2006a). Performance evaluation of a fair fault-tolerant mutual exclusion algorithm. In *Proceedings of the 25th IEEE Symposium on Reliable Distributed Systems (SRDS 2006)*, pages 225–234, Los Alamitos, CA, USA. IEEE Computer Society.

[Sopena et al., 2006b] Sopena, J., Arantes, L., and Sens, P. (2006b). Un algorithme équitable d'exclusion mutuelle tolérant les fautes. In *Actes de la 5ème Conférence Française sur les Systèmes d'Exploitation (CFSE'06)*, pages 97–107, Perpignan, France.

[Sopena et al., 2009b] Sopena, J., Baarir, S., and Legond, F. (2009b). Vérification formelle d'un algorithme générique et hiérarchique d'exclusion mutuelle. *Technique et Science Informatique*, 28.

[Sopena et al., 2007] Sopena, J., Legond, F., Arantes, L., and Sens, P. (2007). A composition approach to mutual exclusion algorithms for grid applications. In *Proceedings of the 36th International Conference on Parallel Processing (ICPP07)*, pages 65–75. IEEE Computer Society.

[Sopena et al., 2008b] Sopena, J., Legond, F., Arantes, L., and Sens, P. (2008b). Composition d'algorithmes d'exclusion mutuelle pour les grilles de calcul. In *Actes de la 6ème Conférence Française sur les Systèmes d'Exploitation (CFSE'08)*, Fribourg, Suisse.

[Suzuki and Kasami, 1985] Suzuki, I. and Kasami, T. (1985). A distributed mutual exclusion algorithm. *ACM Transactions on Computer Systems*, 3(4):344–349.

Chapter 3

Data replication in grid environments

Thi-Mai-Huong Nguyen

Applied Mathematics and Systems Laboratory, Ecole Centrale Paris, Grande Voie des Vignes, 92295 Châtenay-Malabry, France

Frédéric Magoulès

Applied Mathematics and Systems Laboratory, Ecole Centrale Paris, Grande Voie des Vignes, 92295 Châtenay-Malabry, France

3.1 Introduction

Replication is a common technique to improve the performance of data access in distributed systems. In the data grid environment, replication strategies are crucial for reducing data access latency, increasing data locality and availability, and hence improving overall performance in job execution.

In this chapter, we address the replication problem in data grids with the consideration of limited replica storage. We survey existing replication management solutions for data grids and distributed environments including databases, peer-to-peer systems, and web environments. We then discuss the system architecture for the proposed replication strategy called maximize data availability with selective rank (MaxDAR). In our approach, replication decisions are driven by the optimization of the overall level of data availability of the system according to a selective-rank in minimizing the replica's storage costs. Simulation results demonstrate that MaxDAR achieves a better performance in job execution and storage consumption compared with other strategies implemented in OptorSim.

3.2   Data replication

Replication is a well-known technique to improve reliability, load balancing, and performance of the system. In the context of distributed systems, the probability of network failures or data unavailability is higher than in a central system with a single storage resource. In general, data replication can be used to improve system performance by increasing availability of data and fault tolerance. An ideal replication solution should achieve three goals, though: fault tolerance (i.e., data availability), performance (i.e., data locality), and consistency. The goal of replication in each type of environment is slightly different as illustrated in Figure 3.1.

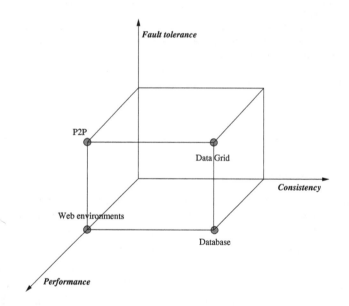

FIGURE 3.1: Focus of replication in each environment.

In this section, some existing approaches to replication management for data grids and distributed environments including databases, peer-to-peer systems, and web environments are shortly discussed. We do not go into detail but want to give an overview of important work in the research community.

3.2.1 Replication in databases

Although replication is not a central emphasis of the database community, interest in dynamic data replications in distributed database management systems continues to grow due to the need to improve scalability, data availability, and fault tolerance. There are many similarities in the replication techniques developed from both the database community and the distributed system community. In [Wiesmann et al., 2000a], a model allowing us to compare and distinguish replication solutions developed for these two perspectives is provided. A key difference of replication solutions in these communities is the nature of replication protocols used [Wiesmann et al., 2000a], [Wiesmann et al., 2000b]. Since the probability of failures in distributed systems is usually higher than database systems, e.g., various hardware and software resources involved might crash, replication in distributed community puts emphasis on fault tolerant systems, i.e., keeping the systems running in case of these failures or data unavailability. On the contrary, the main focus of replication in database systems is to maintain consistency of replicated data and ensure data safety for better application performance. Replication techniques for database systems depend on the update frequency and user access/update patterns on data stored in these databases. If there are more updates in the database, keeping the replicas consistent with each other becomes more costly. In this case, a replication strategy that strictly controls the number of newly created replicas is preferable. Otherwise, wide replication is more profitable as access to data held locally is faster than remote access through wide area network.

The most simple replication protocol in database systems is read-one-write-all (ROWA) protocol, which provides an optimistic degree of data consistency. In ROWA protocol, when a replica is modified all other replicas need to be synchronized and are written at the same time. Since all replicas are supposed to be consistent to each other, a read operation can be satisfied from different replicas. It can select a replica closest to the site of request, and hence benefits from reduced access latency. However, an update operation may negatively affect the performance of the system as locks on replicas have to be agreed among all replicas. In the presence of update performance primary-copy (i.e., master-slave) and multi-master (i.e., update anywhere) are two main replication approaches used in database systems. In the primary-copy approach, the update operation is only permitted on the primary copy of data. Content updates are then propagated to all secondary copies. Depending on the degree of consistency required, different replication protocols can be used, such as quorum-based protocol [Bernstein et al., 1987] and epidemic mechanism [Demers et al., 1987]. Centralized updates at primary copy may become a single point of failure as failures at the master site block update operations. In multi-master approach, many primary copies of the same data item are distributed to multiple sites. The fact that update operations can be concurrently performed to all these copies avoids bottlenecks and single point of failures of primary-copy approach. However, the reconciliation algorithm for ensuring data consistency in multi-master approach is costly due to the potential conflicting updates on primary copies [Martins et al., 2006].

3.2.2   Replication in peer-to-peer systems

Peer-to-peer (P2P) systems are distributed systems that operate without centralized global control in the form of a global registry, global services, global resource management, a global schema, or data repository. In a P2P network, each peer (i.e., participant node) takes both the role of client and server. As a client, it can consume resources offered from other peers and also as a server, it can provide its services for others. P2P systems rely on replication of data on more than one node to achieve a load balance, high data availability and good query latencies, i.e., the number of contacted peers for each data request is reduced. A thorough study and survey of data replication in P2P systems is presented in [Martins et al., 2006]. Replication techniques in P2P systems can be classified into three main categories: passive replication, cache-based replication, and active replication [Androutsellis-Theotokis and Spinellis, 2004].

- Passive replication: When a search is successful, the replication occurs naturally and the requested object is stored at the node which requested it. This is referred to as passive replication. Example of P2P systems that apply passive replication are Napster [Fanning, 2001] and the Gnutella network [Gnutella, 2009].

- Cached-based replication: In this form of replication, the requested object is replicated to every node along the path of the query which requests it. In that way, copies of the data item are cached on all intermediated nodes, improving its availability response latency of future requests. This technique is taken by many systems, such as OceanStore [Kubiatowicz et al., 2000] and Freenet [Clarke et al., 2002].

- Active replication: A peer actively attempts to replicate a data item at other nodes participating in the network. More copies of a data item in the network will generally improve lookup performance, bandwidth, and aggregate network load. Replica management techniques proposed usually differ on the number of replicas (e.g., uniform or proportional) and location of replicas. For example, introspective replica management techniques employed by OceanStore are based on top of extensive caching, whereby traffic is observed and replicas of data items are created to accommodate demand and maintain sufficiently high levels of data redundancy [Rhea et al., 2001].

3.2.3 Replication in web environments

As the Internet has become an indispensable means for content sharing and distribution, the issues of scalability and performance become increasingly important. Caching and replication, being two major techniques used in web environments that address these issues, are becoming a focus of attention in both industrial and academic research communities. Both techniques deal with two important issues: data placement and replacement. Data placement concentrates on selecting location to place a new data, while data replacement decides which data should be removed to make place for a new data. However, caching is performed on the client side in order to improve the access latency and reduce network and server loads, e.g., browser cache, proxy cache, server main memory cache. In contrast, replication is an optimization process at the server side in order to place data close to the request, e.g., web site mirroring. In [Barish and Obraczka, 2000], [Loukopoulos et al., 2002], the authors present an overview of recent research in caching and replication on the Internet.

Different caching techniques have been studied in [Barish and Obraczka, 2000] to improve cache hit ratio and reduce web traffic. While useful and necessary, caching still has some limitations compared to replication. Firstly, many data items are not cacheable but replicable due to their size, whereas data size is a major concern in data grids. Secondly, replication is typically under full control of the service provider, which implies that issues like copyright enforcement, data consistency are much easier to control. Consequently, web cached data may become inconsistent and stale, thereby leading to data integrity problems. Thirdly, servers are free to use replication protocols that would be more efficient for moving replicas between servers or maintaining their consistency, while caches depend heavily on standards.

With the continuous increase of web hosts at large scale, manual placement of a large amount of replicas becomes unfeasible. This has motivated the development of dynamic replication strategies in web environments. The main challenges which have been identified for the implementation of a replicated service on the Internet [Loukopoulos et al., 2002] are: (i) how to assign requests to servers according to some performance criteria; (ii) number and placement of the replica; and (iii) maintaining content consistency in presence of update requests.

Most replication techniques used in web environments follow a primary-copy approach and focus on maintaining consistency of secondary copies. Typically, content updates are propagated from the primary copy to other existing replicas in an asynchronous fashion using different replication protocols, such as quorum-based protocol [Bernstein et al., 1987] and epidemic mechanism [Demers et al., 1987] depending on the needs of the application. In [Pierre et al., 2002], the authors propose optimistic replication techniques, which make use of weak consistency algorithms. Allowing inconsistent access to data is a popular solution to the scalability problem, which comes from the fact that a large amount of replicas are required to be globally synchronized. The inconsistency level is determined based on access/update patterns, as well as the consistency requirements of the application.

One form of replication on the Internet is through content delivery networks (CDNs), such as Akamai [Akamai, 2009] and Digital Island [Digital island, 2009]. A CDN is a set of geographically distributed servers that offer the facilities for hosting web content in order to reduce the load on the origin servers as well as the network traffic. The basic idea used by Akamai, a popular CDN consisting of 16,000 servers across the globe, is to cache the embedded documents that are most used and serve them from the closest CDN server. A content provider can sign up for the service and have its content placed on the content servers. The content is replicated either on-demand when users request it, or it can be replicated beforehand, by pushing the content on the content servers [Kangasharju et al., 2002].

3.2.4   Replication in data grids

Modern supercomputer systems connected through high bandwidth networks have spurred a new class of data-intensive applications. In recent years the data requirements of both scientific and business applications have been dramatically increasing in both volume and scale. The amount of data generated by data-intensive applications continues to grow each year, and the aggregated data volume will reach the exabyte (1 million terabytes) scale by around 2015 [Particle physics data grid, 2009]. Several good examples can be listed such as high energy physics (HEP) experiments [Compact muon solenoid, 2009], and CERN's Large Hadron Collider (LHC) experiments [Large hadron collider, 2009], which processed and produced hundred of terabytes of data. In such applications, data files required might be located in different geographical distributed systems, implying that availability and consistency of data has to be maintained under wide-area environments where network latencies are generally long [Dullmann et al., 2001]. Therefore, there is a great need for an integrated architecture which facilitate the storage, processing, and management of data of both large scale and volume.

Data grid is becoming a promising infrastructure for data storage and execution of data intensive applications, which connects a collection of hundreds of geographically distributed computers and storage resources located in different parts of the world to facilitate sharing of data and resources [Lamehamedi et al., 2003]. Some examples of data grid are the European DataGrid project [European datagrid, 2009], physics data grids [Particle physics data grid, 2009 , GriPhyN, 2009], the LHC computing grid (LCG) project [Large hadron collider, 2006] for handling the massive amount of data produced from the LHC experiments at CERN, and the biomedical informatics research network (BIRN) [Biomedical informatics research network (BIRN), 2005].

Traditionally, grid resources (e.g., computational power, data storage, network bandwidth) are allocated to the jobs by the workload scheduler according to the job requirements, the system load, and specified policies. In a data grid environment, an efficient scheduler must also take into account the location of data required by the jobs, which has, in fact, a significant influence on system performance. The reason for this is that the jobs may take a long time to finish because of a long delay in the fetching of required data files on a high latency storage or just hang due to data unavailability. Data replication, which involves the creation of identical copies of data files and their distribution over various sites, is an important technique to avoid such situations. Appropriate placement of data files at different sites in the system not only reduces the data access time of the jobs, bandwidth consumption, and consequently improves the job turnaround time [Stockinger et al., 2002], [Ranganathan and Foster, 2001], but also increase data availability in many applications [Hoschek et al., 2000], [Ranganathan et al., 2002], [Lei et al., 2008]. Recently there has been a considerable interest in the area of dynamic replication in data grid environments.

Many research works have been carried out to provide basic functionalities of replica management in data grid implementations [Chervenak et al., 2002], [Chervenak et al., 2005], [Gridpp, 2009]. These existing components and services can serve as the basis of the replication services or management framework for grid environments at a higher level to provide an automated replica placement support and a full replica management functionality for achieving better access performance, availability, and security of data. In this section, we overview the existing works in data replication and placement in data grids.

In data-intensive applications where data are usually generated from an instrument at one site and are transferred to other storage sites in data grids through the data replication mechanisms, the data consistency is not a big concern as there are no updates. However, as the application domain of data grid continues to expand, the replication strategies need to address the environments where update requests are frequent. Hence, the replication strategies can be distinguished based on whether they are used for read-only or update requests.

Replication strategies for read-only requests In the context of replication strategies for read-only data environments, replication algorithms can be classified into three main categories.

- Based on replica location: In this model, a replica selection is decided based on a data grid structure and replica location. In [Ranganathan and Foster, 2001], the authors identify five replica strategies (best client, cascading replication, caching, caching plus cascading replica and fast spread) with three different kinds of access patterns (random access, small degree of temporal locality, and small degree of geographical and temporal locality) for a hierarchical data grid. The simulation results show that significant savings in latency and bandwidth can be obtained if the access patterns contain a small degree of geographical locality.

In [Lin et al., 2006], the authors describe a hierarchical tree structure for data grid, where data access requests are generated from the leaf nodes to upper nodes within a range limit. The goal of the replica placement algorithm proposed is to improve the load balancing among the replica servers. The optimal locations for placing the replicas, which are bounded by the minimum number corresponding to the server's workload capacity, are selected based on the usage frequency of users towards a particular data.

In [Park et al., 2003], the authors propose a dynamic replication strategy, called BHR, to reduce data access time by avoiding network congestions in a data grid network. The basic idea of BHR is to take the benefits from a "network-level locality," which represents that required file is located in the site which has broad bandwidth to the site of job execution.

- Based on cost estimation: In the cost estimation model, a replication decision is taken by evaluating the data access gains and the cost of placing a replica of file f at the new location n, which is calculated as:

$$\mathrm{cost}(f, i, j) = f_{\mathrm{transfer}}(\mathrm{bandwidth}_{i,j}, \mathrm{size}_f) + f_{\mathrm{storage}}(f, n) \quad (3.1)$$

In [Lamehamedi et al., 2003], the authors propose a cost model based on runtime bandwidth, replica size, accumulated read/write statistics. The model evaluates the data access gains by creating a replica compared with the costs of creation and maintenance for the replica. A hybrid of tree and ring topologies of nodes was proposed to overlay replicas on the data grid and minimize inter-replica communication cost.

In [Ranganathan et al., 2002], the benefit of creating a new replica is evaluated based on the storage and transfer cost as in equation (3.1). An approach is proposed to create replicas automatically in a decentralized fashion to maintain desired data availability without consuming an undue amount of storage and bandwidth of the system.

- Based on economy policies: Several works [Carman et al., 2002], [Bell et al., 2003] have been proposed based on economy-based policies for replication decisions. The basic idea behind economy-based replication is to apply the economic concepts of market behavior, where data files represent the goods. In this model, the investment, i.e., replication decision, is determined by the difference in cost between the price paid to buy a data file and the expected price, i.e., revenue, if the file is sold in the future. In [Carman et al., 2002], a function that calculates the revenue obtained over a future time period by selling a file F corresponding to the identifier f is defined as:

$$V(f,k,n) = \sum_{i=k}^{k+n} p_i \; \delta(f,f_i) \; \delta(s,s_i) \qquad (3.2)$$

The function $V(f,k,n)$ computes the revenue for the file F corresponding to the identifier f starting from time t_k in the future and taking into account the next n file requests. p_i represents the price paid for the file, which can be calculated based on the equation (3.1). s is the local storage element. δ that represents the incomes of file F returns the value 1 if the arguments are equal and the value 0 if they differ. A replication decision is taken only when the replica with an associated file purchase cost is proved to be potentially beneficial for the system in long term.

A predicting function $E[V(f,k,n),r]$ is defined in [Bell et al., 2002], [Bell et al., 2003], [Capozza et al., 2002] to estimate $V(f,k,n)$ that returns the number of times the file f will be requested in the next period of time considered, based on the latest r file requests. Simulation results realized in OptorSim [Cameron et al., 2004b] present some specific realistic cases where the economic model shows tremendous performance improvements over traditional methods.

In [Lei et al., 2008], the authors propose two new metrics, namely system file missing rate and system bytes missing rate, to evaluate the reliability of the system. An on-line optimizer algorithm (MinDmr) is proposed to maximize the data availability based on these two metrics and the predicting function $E[V(f,k,n),r]$ from [Bell et al., 2003].

Replication models for update-requests In an environment where each replica gets updated frequently, it is desirable to minimize the divergence between the source data and its replicas by a synchronization process. There are two main approaches for a synchronization process to maintain the data consistency: (i) synchronous replication and (ii) asynchronous replication. Synchronous strategy is also known as *eager replication*, while asynchronous is often referred to as *lazy replication*.

- Synchronous: Ideally, replicas need to be kept consistent with the source data at all times. In synchronous replication, updates to any replica should be immediately propagated to all other replicas within the same (distributed) transaction. Synchronous replication enforces mutual consistency of replicas. Although synchronous replication provides strong consistency and a high degree of fault-tolerance, it cannot scale up to large distributed environments like grid environments due to the complexity and cost of the distributed transactions. Synchronous replication strategies, which relies on ROWA or read-one-write-all-available (ROWAA) protocol [Bernstein et al., 1987], are more suitable for database systems. A better solution for data grids that scales up is asynchronous replication.

- Asynchronous: In grid environments, data collections can be very large or frequently updated and network or computational resources may be limited; propagating the updates to all replicas may be infeasible. In certain situations where exact data consistency is infeasible, propagating data updates to a large number of replicas every time a data item replica is modified greatly affects the performance of the entire system. Asynchronous replication is more suitable for this environment where replicas can be updated in different transactions at different sites in an asynchronous fashion. While synchronous replication updates every replica before committing the original update transaction, asynchronous replication commits the original update transaction before updating is completed. According to where the original update transaction is completed, asynchronous replication strategies can follow a primary-copy (i.e., master-slave) or multi-master (i.e., update everywhere) approach. In a primary-copy approach, all the update transactions are forced to be executed and committed on the site holding the primary copy and the site then propagates the updates in various methods. In a multimaster approach, the update transactions are executed and committed on a group of primary copies and then propagated to other replicas. However, conflicting updates at different sites can introduce replica divergence in this approach. Manual or automatic reconciliation processes are used to avoid such situations.

3.3   System architecture

A replication algorithm determines how many replicas are needed, when to create new replicas and where the new replicas, should be placed. To make these decisions, the algorithm needs to choose among alternative plans, the best one based on system parameters. There is no replication strategy that is optimal for all evaluation metrics. In most cases, we have to sacrifice some evaluation metrics in order to make one or other better.

In the early days of distributed systems when network bandwidths were low, decreasing the data access latency and the network bandwidth assumption were the main motivation. As bandwidths increased and communication costs have become relatively cheaper, fault tolerance and availability of data even if one site becomes unavailable have become the new focus.

In this research chapter, we address the replication problem in data grids to improve data availability, data access latency, and bandwidth consumption of the system. We propose a selective-rank replication strategy with the consideration of limited replica storage. In our approach, the focus is to increase the overall level data availability of the system according to a selective-rank, which indicates the importance degree of the file. In fact, the rank of a file can be determined by different metrics, e.g., the degree of popularity of a file depending on specific contexts and circumstances. We propose two newly defined metrics: file popularity and file correlation.

Based on these metrics, we propose a novel algorithm, maximize data availability with selective rank (MaxDAR), that maximizes the overall level data availability of the system while minimizing replica storage cost. Due to storage constraint, some replicas will be deleted when new ones are created at different locations. As a result, MaxDAR evaluates the benefits and losses in overall system level availability before adding any new replica to decide whether it is worth to delete existing files. The replication decisions are made based on several factors, such as file rank according to the popularity or correlation of files, number of replicas, replica size, and file availability benefits. Based on MaxDAR, we introduce three optimizers that distinguish from each other in the selective-rank used. To validate our algorithm, we use the OptorSim simulator [Cameron et al., 2004c , Bell et al., 2002]. Experimental results show our replication optimizers reduce the network traffic greatly and improves the job execution performance effectively.

In data grid, data are regarded as the center of grid architecture because data-intensive applications in the data grid need access to distributed large data sets. We adopt the same view of a data grid structure as that proposed by the European DataGrid project [European datagrid, 2009]. Figure 3.2 depicts the typical system architecture of the data grid.

The data grid consists of a set of sites, and each site has several computing elements (CE), i.e., clusters of machines, which offer computational resources for the jobs. Some sites may have associated storage elements (SE), which provides storage for the replicas. Some isolated SEs might exist in the system in order to improve remote data access performance with each one constructing an independent site by itself.

Jobs are distributed to CEs by the resource broker based on specific scheduling strategies. The required input data must be locally available in the associated SEs before the job begins to execute. In many situations, it may happen that a job running on a CE needs to access data from a remote SE. In the data grid environment there could be replicas of the same file at different sites, each with a different file access cost. In this case, the replica manager containing a replica optimizer is responsible for choosing the best replica in the system and transferring it to the CE performing the job.

The data files are usually replicated at different SEs in the system for effi-

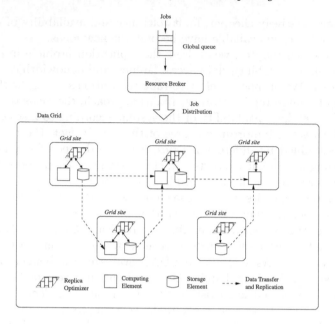

FIGURE 3.2: System architecture of the data grid.

cient access and reducing transmission cost and link latency [Cameron et al., 2004c], [Cameron et al., 2004a] in order to minimize the average job execution time in data grid. The replication operation is independently determined by each site. For every data access request of the job locally running at the site, the replica optimizer determines whether the data should be replicated to local storage and which replicas should be removed if there is not enough space.

3.4 Selective-rank model for a replication system

In a typical data grid, replica optimization is important as efficient job scheduling in the process of maximizing overall job throughput [Cameron et al., 2004c]. Generally, grid users are concerned with the execution time of their submitted jobs and the correctness of the results. Concretely, they are interested in minimizing the total execution time of their jobs and increasing the reliability of execution results. Replicating data files at different sites to achieve data availability as high as possible is an important mechanism for reducing data access time and hence the total job execution time. Therefore, we focus on how to replicate a set of files so as to optimize file availability.

We introduce in this section notations and terminologies to formulate the replication problem. We firstly describe assumptions and system metrics that we use to construct our model. Table 3.1 summarizes the symbols used in our model. Secondly, we introduce how to estimate the availability of files. Finally, we formulate the replication problem based on two main factors, i.e., file availability and its selective-rank.

Parameter	Meaning
M	total number of SEs in the system
N	total number of files in the system
α_i	availability of SE i
ρ_i	availability of a replica of file i
P_i	availability of file i
$P(J_i)$	availability of file for job i
$P_{overall}$	overall level data availability of the system
s_i	size of SE i
$\mathcal{R}_i$	replica set of file i
nr_i	number of replicas of file i, $nr_i = |\mathcal{R}_i|$
$R = [r_{i,j}]$	matrix of replica placement
V_i	popularity of file i
$C(f_i, S_k)$	correlation of file i on site k
$D(f_i, S_k)$	average distance of the file i on the site k to all replicas of other files in the same file set
sc_i	storage cost of file i

Table 3.1: Parameters and their meanings.

3.4.1 Model assumptions

In our model, we assume: (1) a replication system with a fixed population of $\mathcal{S}$ sites connected by a communication network, the number of sites $M = |\mathcal{S}|$; (2) a set of N files in the whole system; (3) each SE i is described by two parameters: storage capacity s_i and online availability α_i, where $i \in [1, M]$.

Storage capacity: each SE i is supposed to offer a certain limited amount of storage space for replication purposes, denoted by s_i.

Online availability: the online availability $\alpha_i \in [0, 1]$ represents the expected probability of time that the SE i is online. When a SE is available, all the replicas it stores are assumed to be available and accessible to other sites in the system.

$$\alpha_i = \frac{MTBF_i}{MTBF_i + MTF_i}$$

Here, $MTBF_i$ represents "mean time between failure" and the MTF_i represents "mean time of failure" of the SE i.

3.4.2   Estimating the availability of files

As the replicas stored on an available site are themselves supposed to be available and accessible to other sites in the system, the availability $\rho_i \in [0, 1]$ of replica of file i on site j is equal to the availability α_j of site j. As a result, a file is available when at least one of its replica or itself is online. Therefore, we can estimate the availability of file i, P_i, according to binomial distribution[1] with the assumption that each replica of file i gives an availability ρ_i, as:

$$P_i(h, nr, \{\rho_i\}) = \sum_{h=1}^{nr} \binom{nr}{h} \overline{\rho}^h (1 - \overline{\rho})^{nr-h} \tag{3.3}$$

where nr represents the number of replicas of file i in the system and $\overline{\rho}$ refers to the average availability of probability set $\{\rho_i\}$.

$$\overline{\rho} = \frac{1}{nr} \sum_{i=1}^{nr} \rho_i \tag{3.4}$$

Suppose that a job i requires access to a file set $\mathcal{F}_i$; the availability of files $P(J_i)$ required for the execution of job i can be computed as:

$$P(J_i) = \prod_{j=1}^{k} P_j, \quad \forall k \in [1, N] \tag{3.5}$$

where $k = |\mathcal{F}_i|$. This approximation in fact estimates the probability required for all k files being available at the moment the job is executed.

3.4.3   Problem definition

The best system data availability results from maximizing equation (3.3) and (3.5). For this purpose, we define the replication matrix $\mathbf{R} = [r_{i,j}]_{M x N}$,

---

[1]The binomial distribution returns the probability $p(k, n, q)$ to obtain k successes performing n independent trials of a certain test when the probability of success of each single trial is q. The binomial distribution is given by the formula:

$$p(k, n, q) = \binom{n}{k} q^k (1 - q)^{n-k}, \quad 0 \le k \le n$$

where $r_{i,j}$ indicates whether a replica of file j is assigned to site i.

$$r_{i,j} = \begin{cases} 1, & \text{if site } i \text{ stores a replica of file } j; \\ 0, & \text{otherwise}; \end{cases}$$

where $i \in [1, M]$ and $j \in [1, N]$. Let r_j refer to the jth column of the replica placement matrix $\mathbf{R}$, which denotes the subset of sites where the file j is replicated. Let nr denote the number of replicas of file j stored in the system:

$$nr_j = \sum_{i=1}^{M} r_{i,j}, \quad \forall j \tag{3.6}$$

Obviously, the total size of all the replicas stored at site i should not exceed its storage capacity s_i:

$$\sum_{j=1}^{N} r_{i,j} \, \text{size}(j) \leq s_i, \quad \forall i \in [1, M] \tag{3.7}$$

where $\text{size}(j)$ is the size of replica j, supposing that a file is associated with a specific rank q, which indicates its importance degree. Applying equation (3.3) and taking into consideration the file rank, we define the overall file availability of the system as follows:

$$P_{overall}(\mathbf{R}) = \frac{\sum\limits_{j=1}^{N} q_j \, P_j(h, nr_j, \{\rho_j\})}{\sum\limits_{j=1}^{N} q_j} \tag{3.8}$$

where $P_j(h, nr_j, \{\rho_j\})$ is the availability of file j, nr_j denotes the number of replicas of file j and can be calculated by equation (3.6), $\{\rho_j\}$ denotes the availability of sites on which a replica of file j is replicated (i.e., $r_{i,j} = 1$).

Given the above equation, we can formulate the data replication problem as to find the assignment of $r_{i,j}$ values in the $\mathbf{R}$ matrix that maximize the availability of files in the system. Our objective then in the design of a replication algorithm is to optimize (i.e., maximize) the overall system level availability $P_{overall}$ in equation (3.8), subject to the storage constraint in equation (3.7). With such a replication algorithm, highly ranked files will receive higher availability. The optimal assignment of the replicas to the appropriate sites is a typical "Knapsack Problem" in consideration that each file replica has a storage cost.

3.5   Selective-rank replication algorithm

The replication algorithm decides which file should be replicated, when to create new replicas and where the new replicas should be placed. To make these decisions, the algorithm needs to choose among alternative plans the best one based on system parameters. In a large-scale and dynamic grid environment, the replication algorithm must also consider changes coming from the underlying environment, e.g., network bandwidth or system workload.

In this research chapter, we propose a selective-rank replication algorithm, which takes into consideration the changes of grid environment, and it automatically creates new replicas for a data file according to its selective-rank while removing old replicas when necessary to improve the performance.

Our algorithm aims to increase the system level data availability according to equation (3.8), based on the selective-rank, which indicates the importance of the file. In fact, the rank of a file can be determined by different metrics depending on specific contexts and circumstances. We propose two newly defined metrics: *file popularity* and *file correlation*, which can be applied in the equation (3.8) to enhance the overall data availability.

3.5.1   Popularity of files

In data grid, the popularity of file could be thought of as the number of times files are requested in the future by grid jobs. In order to reduce the access time and due to storage constraint, it is more beneficial to replicate frequently accessed files, i.e., "hot" files, than less used ones, i.e., "cold" files. Based on the file access history stored at each site, the number of times it will be accessed in the future could be predicted.

In this chapter, we use the predicting functions given in [Bell et al., 2002], [Bell et al., 2003], [Capozza et al., 2002] to evaluate the popularity of file i, V_i, which are:

- Binomial prediction: V_i is predicted based on file access history using a binomial distribution.

- Zipf prediction: V_i is predicted based on file access history using a Zipf distribution.[2]

3.5.2   Correlation of files

The concept of the correlated degree of files derives from the fact that in data grid some files have a high probability to be used with a specific set

---

[2]A Zipf-like distribution is defined as $P_n \propto n^{-\alpha}$, where P_n is the frequency of occurrence of the nth ranked item and $\alpha \leq 1$ (a pure Zipf distribution would have $\alpha = 1$).

of files. For example, data files containing protein structure information are usually used in biology experiments together with other files in the same discipline rather than be used in high energy physics experiments.

The basic idea behind introducing this concept is that the highly correlated files, i.e., the files which are most likely to be requested by the same job will be gathered into a region made up of relatively close sites, so that jobs executed in that region can take advantage of the reduced cost to transfer required data files to the executing site. The data access latency for the job processing and, hence, the job execution performance will be improved.

Suppose each site keeps a list of file sets requested by different jobs. For example: $\mathcal{F}(f_1, f_2, f_3, f_4), \mathcal{F}(f_5, f_6), \mathcal{F}(f_7, f_8, f_9)$. We define that data files which are "correlated" to each other are the ones belonging to the same file set required by a job. The file correlation $C(f_i, S_k)$ reflects the correlation degree of the file i on the site k with other files in the same file set. A file i is supposed to be close to other files in the same file set when it is located at a site k that offers higher $C(f_i, S_k)$, which can be calculated as:

$$C(f_i, S_k) = \frac{\text{size}(i)}{D(f_i, S_k)} \tag{3.9}$$

where $D(f_i, S_k)$ evaluates the average distance of the file i on the site k to all replicas of other files in the same file set and is computed as:

$$D(f_i, S_k) = \frac{1}{|\mathcal{F}|} \sum_{f_i, f_j \in \mathcal{F}} \frac{1}{c_{i,j} |\mathcal{R}_j|} \sum_{v \in \mathcal{R}_j \ v \neq S_k} \frac{\text{size}(v)}{\text{bandwidth}(S_v, S_k)} \tag{3.10}$$

where $\mathcal{R}_j$ is the replica set of f_j, which belongs to the same file set $\mathcal{F}$ of f_i. S_v is the site containing the replica v of f_j. In this estimation, all replica located on the site S_k are excluded. In the above equation, $c_{i,j}$ is given by $min\{c(f_i, f_j), c(f_j, f_i)\}$ where $c(f_i, f_j) \in [0, 1]$ denotes the probability of two files f_i and f_j occurring in the same file set (i.e., requested by the same job) and can be computed as:

$$c(f_i, f_j) = \frac{\text{count}(f_i, f_j)}{t(f_i)} \tag{3.11}$$

where $count(f_i, f_j)$ denotes the number of distinct files set that both f_i and f_j belong to, and $t(f_i)$ represents the total number of files set that f_i belongs to.

3.5.3 MaxDAR optimizer algorithm

We propose an original algorithm called MaxDAR (maximize data availability with selective rank) optimizer algorithm as shown in Algorithm 3.5.1. The main task of the MaxDAR optimizer is to determine whether or not a new replica is created based on the benefits received in the overall data availability of the system. The goal of MaxDAR is to increase the availability of the most highly ranked files in order to avoid situations in which storage resources are wasted by unimportant files. In this way, the highly ranked, or important files have a high possibility to have greater storage resources delivered to them – hence achieve a higher availability according to their requirements. This strategy makes sense especially in the context of limited storage resources.

Due to storage constraints there must be an efficient mechanism to delete the existing files from the sites for replacement. The replacement strategy in [Ranganathan and Foster, 2001] proposes to delete the most unpopular files (i.e., LFU), once the storage space of the site is exhausted. The other popular strategy for replica replacement is to delete the least recently used files [Bell et al., 2003] (i.e., LRU).

In our algorithm, existing files are deleted to gain space for the new replica in case there is not enough free space. The candidate files to be replaced will be selected based on their storage cost. We introduce the storage cost of file i as:

$$sc_i = \frac{nr_i \text{ size}(i)}{P_i \ q_i} \tag{3.12}$$

where nr_i denotes the number of replicas and size(i) represents the size, P_i reflects the availability, q_i denotes the rank of the file i.

As shown in Algorithm 3.5.1, if the required file does not locally exist in the SE, it will be fetched from other sites (line 1-3). Then, if the free storage space is large enough to store the requested file, the replication of the file will always take place (line 4-5). Otherwise, a set of candidate files C to be deleted for the new replica will be selected according to their storage cost (line 7-12). Since the goal of MaxDAR is to maximize the system level data availability according to a selective-rank, the replication benefits are required to be greater than the replacement loss (line 13-18) for the replication to take place. The replacement loss is evaluated by the sum of the availability loss of the selected candidate files according to a selective-rank (line 13). The P_i and P_i^{new} is the availability of file i before and after the replication and is calculated by equation (3.3). It should be noted that the algorithm will ignore the master files, i.e., primary copies of the data file, and pinned files, i.e., files that are being accessed by jobs, in the selection of candidate files for the replacement.

Based on MaxDAR algorithm, we propose three variant optimizers:

- MaxDAR-Pb: Files are ranked according to their popularity. The predicting function for the number of file access in the future is calculated by a binomial distribution, $q_i = V_i$. This optimizer focuses on replicating the frequently accessed files.

- MaxDAR-Pz: Files are ranked according to their popularity. The predicting function for the number of file access in the future is calculated by a Zipf distribution, $q_i = V_i$. This optimizer focuses on replicating the frequently accessed files.

Algorithm 3.5.1 MaxDAR (Maximize Data Availability with selective Rank)

1: **if** needed file $f_i \not\exists$ in the site **then**
2:     Get f_i from other sites
3: **end if**
4: **if** free space in SE $> size(i)$ **then**
5:     Store f_i in this SE
6: **else**
7:     Sort files in the SE in the descending order of storage cost (equation (3.12))
8:     $\mathcal{C} \leftarrow \{\}$
9:     **while** $\sum_{f_k \in \mathcal{C}} size(k) < size(i)$ **do**
10:         Pop the first file $f_{candidate}$ off the sorted list
11:         $\mathcal{C} = \mathcal{C} \cup \{f_{candidate}\}$
12:     **end while**
13:     $loss = \sum_{f_k \in \mathcal{C}} P_k \, q_k$
14:     $benefit = P_i^{new} \, q_i$
15:     **if** $benefit > loss$ **then**
16:         Delete all the file in $\mathcal{C}$
17:         Store f_i
18:     **end if**
19: **end if**

- MaxDAR-C: Files are ranked according to their correlation with other files located near the site where the replication is considered, $q_i = C(f_i, S_k)$. The replication decision of file i on the site k is evaluated by the value of $C(f_i, S_k)$ (equation (3.9)). As indicated in Section 3.5.2, this optimizer aims to replicate in priority the files that are likely to be requested in the same job so that when the job is executed nearby, the cost for file access will be reduced. The running times and hence the costs of running jobs are also reduced.

In the next section, we present the performance evaluation of our proposed MaxDAR optimizers with simulation experiments using the grid simulation tool OptorSim [Cameron et al., 2004b].

3.6   Evaluation

We evaluated our proposed MaxDAR optimizers using the OptorSim v2.0.1, which was developed by the European DataGrid project [European datagrid,

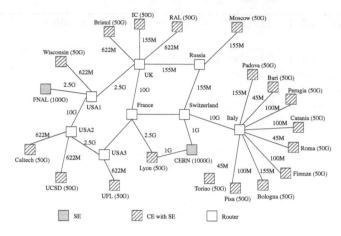

FIGURE 3.3: Grid topology in the simulation.

2009] for evaluating both the file access optimization and data replication strategies. OptorSim simulates the architecture shown in Figure 3.2, which consists of several components: computing elements (CEs), storage element (SEs), resource broker, and replica optimizer. Simulated jobs are distributed to the optimal CEs across the grid according to scheduling algorithm used by the resource broker. As the CE requests the file set for each job, the replica optimizer decides whether or not to replicate each file according to the benefit gains from the replication.

We have implemented three MaxDAR optimizers as three new replica optimizers in OptorSim. We first discuss the simulation's configuration, followed by the results.

Parameter	Meaning	Value
M	number of sites	20
N	number of files	97
$size_i, i \in [1, N]$	size of files	1G-5G
$s_i, i \in [1, M]$	SE capacity	50G-1000G
$\alpha_i, i \in [1, M]$	SE availability	0.8-1.0
$|\mathcal{F}_i|$	number of files accessed by a job i	3 - 59
-	number of jobs	200-1000
-	network bandwidth	45M-10G/s

Table 3.2: Parameter settings for the simulation.

3.6.1 Grid configuration

The grid topology used in the simulation is adopted from CMS testbed [GriPhyN, 2009], [Holtman, 2001], which has the resources and network bandwidths between the sites as shown in Figure 3.3. It consists of 8 routers and 20 grid sites in Europe and USA [Cameron et al., 2004c]. We utilize the default settings of the OptorSim and modified the topology to meet our needs. At the beginning of the simulation, all the master files are placed at the CERN site. Initially, there are 97 files in the grid and the storage capacity of SEs ranges from 50 GB to 1000 GB. CERN was allocated huge SEs of 1000 GB capacity and no CEs. Every other site excluding FNAL, which was allocated 100 GB, was given 50 GB of storage and a CE with one worker node.

We are interested in how the replication algorithm performs under different parameters, such as the total number of jobs to be executed, the file size, and the data availability at each SE. These parameters are summarized in Table 3.2. It should be noted that our simulation does not take into account the job execution, i.e., the processing of the files by the CEs, but only the file transmission time required by the job.

3.6.2 Experimental results

In this section we present the simulation results. The objective of these simulation experiments is to evaluate the efficiency of file distribution in the context of limited storage and the effect of our replication algorithm on the job execution time and the overall data availability of the system. In the experiments, we compute:

- Job execution time: This is defined as the total time to execute all the jobs, divided by the number of jobs completed.

- Overall file availability of the system $P_{overall}$: This is used to evaluate the quality of data replication at the overall data availability goal based on their selective-ranks, calculated by equation (3.8).

The experiments are performed on a laptop IBM 2.8 G CPU and 1 G RAM. For each simulation setup, we perform it several times and take the average results. For all the experiments except those in Section 3.6.2.3, we simulate 1,000 submitted jobs.

3.6.2.1 Effects of access pattern and scheduling strategies

We first evaluate the performance of three replica optimizers, i.e., MaxDAR-Pb, MaxDAR-Pz, and MaxDAR-C, based on the MaxDAR algorithm in terms of the job execution time. We recall that in MaxDAR-Pb and MaxDAR-Pz optimizer, the predicting functions of OptorSim, whose in-depth discussion is given in [Bell et al., 2002], [Bell et al., 2003], [Capozza et al., 2002], are utilized for the estimation of file popularity.

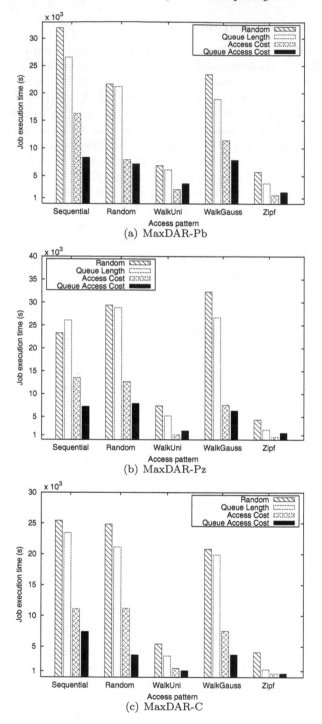

FIGURE 3.4: Job execution time of three optimizers for various scheduling strategies and access patterns.

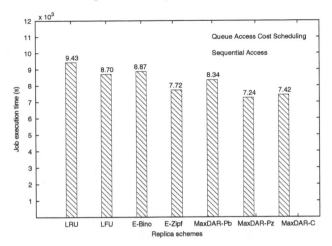

FIGURE 3.5: Job execution time for different replication schemes with sequential access and queue access cost scheduling.

Figure 3.4 shows the results of the job execution time for three optimizers. In this experiment, we study the impacts of the scheduling strategies used by the resource broker and the data access pattern on the performance of three optimizers. We consider four scheduling strategies: random, queue length, access cost, and queue access cost, together with five access patterns: sequential, random, random walk unitary, random walk gaussian, and random walk Zipf. The random scheduler randomly chooses a CE on the grid for a job execution. The queue length scheduler chooses a CE with the shortest job queue for a job execution. The access cost scheduler distributes the submitted jobs to a CE with the lowest data access cost. Lastly, the queue access cost scheduler chooses a CE with the smallest combination of access cost and queuing cost for a job execution.

In general, the three optimizers show very similar performance for each scheduling strategy and access pattern. For each optimizer, the job execution time for random walk unitary and random walk Zipf is approximately half that of other access patterns.

The scheduling strategies random and queue length have the longest job execution time as they do not consider the data location in the job distribution process. We achieved the lowest job execution time when the scheduling strategy queue access cost was utilized. In particular, the MaxDAR-C has the best performance as this scheduling strategy tends to schedule jobs close to the location of the data, while MaxDAR-C replicated the correlated files which were close to each other.

Figure 3.5 shows the comparison of job execution time between our three optimizers and other replication schemes in OptorSim. This experiment was

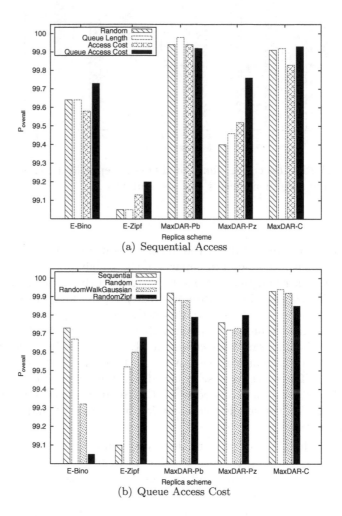

(a) Sequential Access

(b) Queue Access Cost

FIGURE 3.6: The effect of access patterns and scheduling strategies on overall file availability of the system $P_{overall}$.

performed with the sequential access pattern and queue access cost scheduling. Overall, the MaxDAR optimizers have a lower job execution time compared to the LRU, LFU, E-Bino, and E-Zipf replication schemes. Both MaxDAR-Pb and MaxDAR-Pz perform better than E-Bino and E-Zipf respectively. LRU, LFU, and E-Bino have a longer job execution time than all of the MaxDAR optimizers. Although E-Zipf has a shorter execution time than MaxDAR-Pb optimizer, it still has a longer execution time comparing to MaxDAR-Pz.

In the next experiment, we study the MaxDAR optimizers in terms of overall level data availability $P_{overall}$ of the system. Figure 3.6 shows the comparison of $P_{overall}$ between five replication schemes for different scheduling strategies and access patterns. For this experiment, we assume the data availability at each SE is 80%. Figure 3.6(a) and Figure 3.6(b) show the impacts of scheduling strategies and access patterns respectively on $P_{overall}$.

It can be observed from Figure 3.6(a) and Figure 3.6(b) that MaxDAR optimizers achieve better performance than E-Bino and E-Zipf for all scheduling strategies and access patterns. This is predictable because MaxDAR optimizers focus on improving the overall data availability of the system. In the two economical models, the data replication takes place only if this is considered beneficial for the SE in terms of data access cost without considering the data availability gain. As the Zipf based predicting function is not accurate for the sequential access, the replication optimizers based on Zipf distribution, i.e., E-Zipf and MaxDAR-Pz, achieve lower $P_{overall}$ for this access pattern as shown in Figure 3.6(a).

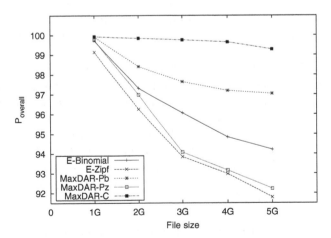

FIGURE 3.7: $P_{overall}$ with sequential access pattern, Queue Access Cost scheduling when varying file size.

Contrary to E-Bino, which has better $P_{overall}$ in sequential access pattern using queue access cost scheduling, E-Zipf achieves better $P_{overall}$ in random access patterns as illustrated in Figure 3.6(b).

3.6.2.2 Performance with different file size

In the next experiment, we are interested in the file size effect on $P_{overall}$. The parameter to be varied was the size of file; the number of submitted jobs is still 1,000. The following size values: 1 G, 2 G, 3 G, 4 G, and 5 G with the sequential access pattern and queue access cost scheduling are utilized. We note that all files still have the same size.

As shown in Figure 3.7, the larger the file size, the smaller the $P_{overall}$. This is predictable because with a limited storage constraint increasing the file size leads to decreasing the number of copies of each file, which in turn decreases the $P_{overall}$. We can see that MaxDAR-C performs best with all file size; MaxDAR-Pb and MaxDAR-Pz perform better than E-Bino and E-Zipf respectively. As expected, Zipf-based optimizers have the worst performance because the Zipf-based predicting functions give an inaccurate prediction of the file value for sequential access pattern.

3.6.2.3 Performance with a different job load

We now examine the scalability of our MaxDAR optimizers by varying the number of submitted jobs. The number of created replicas and job execution time for five optimizers is shown in Figure 3.8(a) and Figure 3.8(b).

As shown in Figure 3.8(a), MaxDAR optimizers make efficient use of the storage resources and network bandwidth with lower number of created replicas. The reason is that E-Bino and E-Zipf try to replicate the most popular files to local storage, which leads to a linear increase in the number of created replicas for the number of executed jobs. On the contrary, MaxDAR optimizers replicate files by considering the overall data availability benefit gain. As the number of sites is fixed, the overall data availability will have an upper boundary; therefore increasing the number of submitted jobs does not affect the number of created replicas.

As illustrated in Figure 3.8(b), all five optimizers produce a similar job execution time. However, as the number of jobs submitted is increased, the performance of the MaxDAR optimizers is improved. Their job execution time gets slightly faster than E-Bino and E-Zipf.

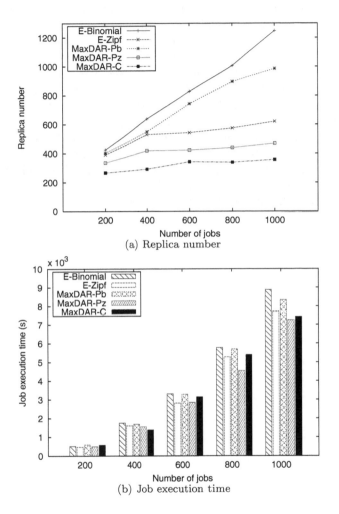

(a) Replica number

(b) Job execution time

FIGURE 3.8: Replication number and job execution time when varying the number of submitted jobs.

3.7   Concluding remarks

In this chapter, we address the issue of providing an efficient data access in data grid environment. We assume that data files are not of equal importance and they can be classified according to their degree of importance, called *selective-rank*. We formulate the replication problem as an optimization problem where the aim is to maximize the overall data availability of the system according to a selective-rank with storage constraints. The proposed replication algorithm maximize data availability with selective rank (MaxDAR) focuses on improving data availability with differentiated replication. The storage cost is utilized for the replica replacement strategy. Based on the MaxDAR algorithm, three variant optimizers have been developed.

The performance of the MaxDAR optimizers is evaluated by simulations in OptorSim. Our simulation results show that the MaxDAR optimizers achieved better performance than other replica schemes of OptorSim whilst also assuring the good overall data availability of systems.

3.8 References

[Akamai, 2009] Akamai (2009). Documentation. Available online at: `http://www.akamai.com` (accessed May 1, 2009).

[Androutsellis-Theotokis and Spinellis, 2004] Androutsellis-Theotokis, S. and Spinellis, D. (2004). A survey of peer-to-peer content distribution technologies. *ACM Computing Survey*, 36(4):335–371.

[Barish and Obraczka, 2000] Barish, G. and Obraczka, K. (2000). World wide web caching: trends and techniques. *IEEE Communications Magazine*, pages 178–185.

[Bell et al., 2002] Bell, W. H., Cameron, D. G., Capozza, L., Millar, A. P., Stockinger, K., and Zini, F. (2002). Simulation of dynamic grid replication strategies in OptorSim. In *Proceedings of the 3rd International Workshop on Grid Computing (GRID'02)*, pages 46–57, London, UK. Springer-Verlag.

[Bell et al., 2003] Bell, W. H., Cameron, D. G., Carvajal-Schiaffino, R., Millar, A. P., Stockinger, K., and Zini, F. (2003). Evaluation of an economy-based file replication strategy for a data grid. In *Proceedings of the 3rd International Symposium on Cluster Computing and the Grid (CCGRID'03)*, page 661, Washington, DC, USA. IEEE Computer Society.

[Bernstein et al., 1987] Bernstein, P. A., Hadzilacos, V., and Goodman, N. (1987). *Concurrency control and recovery in database systems.* Addison-Wesley.

[Biomedical informatics research network (BIRN), 2005] Biomedical informatics research network (BIRN) (2005). Documentation. Available online at: `http://www.nbirn.net/` (accessed May 1, 2009).

[Cameron et al., 2004a] Cameron, D., Casey, J., Guy, L., Kunszt, P., Lemaitre, S., Mccance, G., Stockinger, H., Stockinger, K., Andronico, G., Bell, W., Ben-Akiva, I., Bosio, D., Chytracek, R., Domenici, A., Donno, F., Hoschek, W., Laure, E., Lucio, L., Millar, P., Salconi, L., Segal, B., and Silander, M. (2004a). Replica management in the European data grid project. *Journal of Grid Computing*, 2(4):341–351.

[Cameron et al., 2004b] Cameron, D., Millar, A. P., and Nicholson, C. (2004b). OptorSim: a simulation tool for scheduling and replica optimisation in data grids. In *Proceedings of Computing in High Energy Physics (CHEP 2004)*, Interlaken, Switzerland.

[Cameron et al., 2004c] Cameron, D., Millar, A. P., Nicholson, C., Carvajal-Schiaffino, R., Stockinger, K., and Zini, F. (2004c). Analysis of scheduling and replica optimisation strategies for data grids using OptorSim. *Journal of Grid Computing*, 2(1):57–69.

[Capozza et al., 2002] Capozza, L., Stockinger, K., and Zini, F. (2002). Preliminary evaluation of revenue prediction functions for economically-effective file replication. Technical Report DataGrid-02-TED-020724, CERN, Geneva, Switzerland.

[Carman et al., 2002] Carman, M., Zini, F., Serafini, L., and Stockinger, K. (2002). Towards an economy-based optimisation of file access and replication on a data grid. In *Proceedings of the 2nd IEEE/ACM International Symposium on Cluster Computing and the Grid (CCGRID'02)*, page 340, Washington, DC, USA. IEEE Computer Society.

[Chervenak et al., 2005] Chervenak, A., Schuler, R., Kesselman, C., Koranda, S., and Moe, B. (2005). Wide area data replication for scientific collaborations. In *Proceedings of the 6th IEEE/ACM International Workshop on Grid Computing (GRID'05)*, pages 1–8, Washington, DC, USA. IEEE Computer Society.

[Chervenak et al., 2002] Chervenak, A. L., Deelman, E., Foster, I. T., Guy, L., Hoschek, W., Iamnitchi, A., Kesselman, C., Kunszt, P. Z., Ripeanu, M., Schwartzkopf, R., Stockinger, H., Stockinger, K., and Tierney, B. (2002). Giggle: a framework for constructing scalable replica location services. In *Proceedings of the 2002 ACM/IEEE conference on Supercomputing*, pages 1–17, Baltimore, Maryland, USA.

[Clarke et al., 2002] Clarke, I., Miller, S. G., Hong, T. W., Sandberg, O., and Wiley, B. (2002). Protecting free expression online with Freenet. *IEEE Internet Computing*, 6(1):40–49.

[Compact muon solenoid, 2009] Compact muon solenoid (2009). Project documentation. Available online at: `http://cmsinfo.cern.ch/` (accessed May 1, 2009).

[Demers et al., 1987] Demers, A., Greene, D., Hauser, C., Irish, W., Larson, J., Shenker, S., Sturgis, H., Swinehart, D., and Terry, D. (1987). Epidemic algorithms for replicated database maintenance. In *Proceedings of the 6th annual ACM Symposium on Principles of Distributed Computing (PODC'87)*, pages 1–12, New York, NY, USA. ACM Press.

[Digital island, 2009] Digital island (2009). Project documentation. Available online at: `http://www.digisle.com` (accessed May 1, 2009).

[Dullmann et al., 2001] Dullmann, D., Hoschek, W., Jaen-Martinez, J., Segal, B., Stockinger, H., Stockinger, K., and Samar, A. (2001). Models for replica synchronisation and consistency in a data grid. *High Performance on Distributed Computing*, 0:67.

[European datagrid, 2009] European datagrid (2009). Project documentation. Available online at: `http://www.edg.org` (accessed May 1, 2009).

[Fanning, 2001] Fanning, S. (2001). Napster. Available online at: `http://www.napster.com` (accessed May 1, 2009).

[Gnutella, 2009] Gnutella (2009). Documentation. Available online at: `http://www.gnutella.com` (accessed May 1, 2009).

[Gridpp, 2009] Gridpp (2009). edg-replica-manager 1.0. Available online at: `http://www.gridpp.ac.uk/wiki/EDG_Replica_Manager` (accessed May 1, 2009).

[GriPhyN, 2009] GriPhyN (2009). Grid physics network website project documentation. Available online at: `http://www.griphyn.org/` (accessed May 1, 2009).

[Holtman, 2001] Holtman, K. (2001). The compact muon solenoid (CMS) experiment note: data grid system overview and requirements. Technical Report 2001/037, CERN, Geneva, Switzerland.

[Hoschek et al., 2000] Hoschek, W., Jean-Martinez, J., Samar, A., Stockinger, H., and Stockinger, K. (2000). Data management in an international data grid project. In *Proceedings of the 1st IEEE/ACM International Workshop on Grid Computing (Grid'00)*, volume 1971, pages 77–90, Bangalore, India. Springer-Verlag.

[Kangasharju et al., 2002] Kangasharju, J., Roberts, J. W., and Ross, K. W. (2002). Object replication strategies in content distribution networks. *Computer Communications*, 25(4):367–383.

[Kubiatowicz et al., 2000] Kubiatowicz, J., Bindel, D., Chen, Y., Eaton, P., Geels, D., Gummadi, R., Rhea, S., Weatherspoon, H., Weimer, W., Wells, C., and Zhao, B. (2000). OceanStore: an architecture for global-scale persistent storage. In *Proceedings of the 9th International Conference on Architectural Support for Programming Languages and Operating Systems (ASPLOS-IX)*, pages 190–201, New York, NY, USA. ACM Press.

[Lamehamedi et al., 2003] Lamehamedi, H., Shentu, Z., Szymanski, B., and Deelman, E. (2003). Simulation of dynamic data replication strategies in data grids. In *Proceedings of the 17th International Symposium on Parallel and Distributed Processing (IPDPS'03)*, page 100, Washington, DC, USA. IEEE Computer Society.

[Large hadron collider, 2006] Large hadron collider (2006). Computing grid project. Available online at: `http://lcg.web.cern.ch/LCG` (accessed May 1, 2009).

[Large hadron collider, 2009] Large hadron collider (2009). Documentation. Available online at: http://lhc.web.cern.ch/lhc/ (accessed May 1, 2009).

[Lei et al., 2008] Lei, M., Vrbsky, S. V., and Hong, X. (2008). An on-line replication strategy to increase availability in data grids. *Future Generation Computer Systems*, 24(2):85–98.

[Lin et al., 2006] Lin, Y.-F., Liu, P., and Wu, J.-J. (2006). Optimal placement of replicas in data grid environments with locality assurance. In *Proceedings of the 12th International Conference on Parallel and Distributed Systems (ICPADS'06)*, pages 465–474, Washington, DC, USA. IEEE Computer Society.

[Loukopoulos et al., 2002] Loukopoulos, T., Ahmad, I., and Papadias, D. (2002). An overview of data replication on the internet. In *Proceedings of the International Symposium on Parallel Architectures, Algorithms and Networks (ISPAN'02)*, pages 31–37, Washington, DC, USA. IEEE Computer Society.

[Martins et al., 2006] Martins, V., Pacitti, E., and Valduriez, P. (2006). Survey of data replication in P2P systems. Technical report, INRIA.

[Park et al., 2003] Park, S.-M., Kim, J.-H., Ko, Y.-B., and Yoon, W.-S. (2003). Dynamic data grid replication strategy based on internet hierarchy. In *Proceedings of the 2nd International Workshop on Grid and Cooperative Computing (GCC'2003)*, pages 838–846, Shanghai, China. Springer-Verlag.

[Particle physics data grid, 2009] Particle physics data grid (2009). Documentation. Available online at: http://www.ppdg.net/ (accessed May 1, 2009).

[Pierre et al., 2002] Pierre, G., van Steen, M., and Tanenbaum, A. S. (2002). Dynamically selecting optimal distribution strategies for web documents. *IEEE Transactions on Computers*, 51(6):637–651.

[Ranganathan and Foster, 2001] Ranganathan, K. and Foster, I. (2001). Identifying dynamic replication strategies for a high-performance data grid. In *Proceedings of the 2nd International Workshop on Grid Computing (GRID'01)*, pages 75–86, London, UK. Springer-Verlag.

[Ranganathan et al., 2002] Ranganathan, K., Iamnitchi, A., and Foster, I. (2002). Improving data availability through dynamic model-driven replication in large peer-to-peer communities. In *Proceedings of the 2nd IEEE/ACM International Symposium on Cluster Computing and the Grid (CCGRID '02)*, page 376, Washington, DC, USA. IEEE Computer Society.

[Rhea et al., 2001] Rhea, S., Wells, C., Eaton, P., Geels, D., Zhao, B., Weatherspoon, H., and Kubiatowicz, J. (2001). Maintenance-free global data storage. *IEEE Internet Computing*, 5(5):40–49.

[Stockinger et al., 2002] Stockinger, H., Samar, A., Holtman, K., Allcock, B., Foster, I., and Tierney, B. (2002). File and object replication in data grids. *Cluster Computing*, 5(3):305–314.

[Wiesmann et al., 2000a] Wiesmann, M., Pedone, F., Schiper, A., Kemme, B., and Alonso, G. (2000a). Understanding replication in databases and distributed systems. In *Proceedings of 20th International Conference on Distributed Computing Systems (ICDCS'2000)*, Taipei, Taiwan, R.O.C. IEEE Computer Society.

[Wiesmann et al., 2000b] Wiesmann, M., Schiper, A., Pedone, F., Kemme, B., and Alonso, G. (2000b). Database replication techniques: a three parameter classification. In *Proceedings of the 19th IEEE Symposium on Reliable Distributed Systems (SRDS'00)*, page 206, Washington, DC, USA. IEEE Computer Society.

Chapter 4

Data management in grids

Jean-Marc Pierson

IRIT, Université Paul Sabatier, 118 route de Narbonne, 31062 Toulouse Cedex 9, France

4.1 Introduction

As it was sketched in previous chapters, Grids have been so far very successful to handle at large scale a lot of computing intensive applications in domains like high energy physics, weather forecast, genetics, ... Most of these applications deal with data, either as inputs, outputs or both of their processing. Some of them are dealing with gigabytes, terabytes or even petabytes. In 2006, the digital data production of the humanity reached 161 exabytes. Grids do not deal yet with these numbers, but are each day getting closer. For instance, the Large Hadron Collider at CERN [1] produces 15 petabytes of data each year starting in 2008.

This chapter will introduce Data Management in Grids, and not Grid Data Management. In our understanding the difference is in the integration of the data management in Grids. Studying Grid Data Management would mean describing a running middleware stack, expressing the links between the available services, performances and constraints. We rather plan to raise in this chapter some problems of the data management in Grids, explaining our point using several models and technical implementations in different Grids. In our understanding an absolute and universal Grid does not make

---

[1] lcg.web.cern.ch/LCG.

sense. We will keep the data management as the first concern, while Grids will be the context of the study. Heterogeneity at the software and middleware levels make it impossible to provide an unique answer to the problem of data management in Grids.

As expressed by R. Moore at the Open Grid Forum 22 in February 2008, Data Management in Grids covers a large set of different aspects and interests. In *Data Grids*, the objective is to share scientific data, to organize data as collections so that they can be easily managed as a set. *Digital libraries* are interested in publishing data to the general public: They must support browsing and discovery. Some data must be stored in *persistent archives*, which must take into account the authenticity and integrity of the data and the technology evolution so that data can still be readable in centuries. They must mitigate data loss by means for instance of data replication and federation of catalogs. *Real time sensor systems* produce data on the fly. The problem is then to federate sensor data and to integrate these data across sensor streams. These examples show the difficulty to design a generic infrastructure that covers all the requirements at once.

Transparency is one of the pillars of the Grid. Data management in Grids is no exception: Whatever is the hardware, software, middleware and network environment, data must be accessible in a seamless way. Naming constraints, physical locations, physical hardware components, access methods and potential processing units must not interfere with the provision and management of data.

The scalability issue is of prime importance. The scalability issue has to be handled in terms of the size of data pieces, of number of resources (data pieces, storage nodes) and in terms of number of clients. While Grids can aggregate small scale resources and data, it can also grow up dramatically: As for an example, the EGEE [2] infrastructure gathers 72000 CPUs and a total storage capacity of 20 Petabytes in disk, not counting the tapes in Massive Mass Storage. Data transfer is estimated at 2.3 PBytes per month with an average throughput of 860 Mb/s. The number of registered users in EGEE III is 13800.

Commonly used data in Grids are of many different modalities, depending on the application area and the processing services. It spans from text to video, from raw data to annotated rich data, from structured to fully structured data. For instance, in the high energy physics case, most of the files manipulated are filled with numbers while medical images are more common in the life science cases.

Another point of heterogeneity concerns the data sources or sinks that are used. They can be either raw files, on the fly data, data stored in semi-structured files (like XML), in files systems or in Data Bases Management Systems. Sometimes, one application will use many different modalities, com-

---

[2]www.eu-egee.org.

ing from many different data sources. For instance, the medical grids often deal with MRI or PET images, raw files coming from biogenetic, structured tables in patient databases and raw text files from doctor prescription, expertise and analysis of the situation of the patient. All the available data usually lead from data to information to knowledge.

The heterogeneity appears also at the level of the security administration of the data. The needs are various and depends again on the application area. Security is not much a concern when raw data are considered. But semantically rich data (like medical data from a patient) or enriched data coming from a computerized processing (for instance results of a fluid dynamic algorithm on a plane structure or drug discovery) deserve more attention. Confidentiality and integrity of data must sometimes be ensured at a high level, without compromising their performance or their usability.

Grids suffer from a potential high dynamic of the resources (storage, network, cpus) and users. It is mandatory that the access to data is not constrained by this dynamic. Thus data management must take into account the dynamic in order to guarantee fault tolerance and best performances.

Finally, the coordination of resources available on the Grids must also benefit to and from data management. Indeed, the data management must be tightly coupled with the coordination of the resources. For instance, gathering separated storage disks into a virtual storage space increases the storage capacity of individual entities of the Grid when the coordination and cooperation of these entities in a seamless way is achieved.

The rest of the chapter is organized as follows: In Section 4.3 we will situate data management in Grids in the distributed systems microcosm. In Section 4.4 we will explain links between data management in Grids and the other services one can typically find in Grids. Section 4.5 gives details in the problems related to data management in Grids and gives some hints for solutions. But let us first examine quickly why data are considered coming from data sources.

4.2   From data sources to databases . . . to data sources

It can appear peculiar to speak about databases and data sources. Nevertheless, we believe that this discussion has its place in this chapter. Indeed, a question that should be mentioned is why do we need new tools to access data, while databases have been designed since several decades ago to handle especially this problem.

In former times, before databases were invented, data were stored in flat files, and manual tools and scripts were implemented to handle the access, replication, security related to these data. Then came the time for databases, that provided a level of abstraction above the raw data, with the main benefits of performances and organization optimizations. Data are structured, access to these data are controlled by ad-hoc procedures in data bases management systems, which are in charge of handling efficiently and securely the data pieces. Moreover while data sources contain generally raw data, databases preserve raw data together with data coming from additional process, annotations, etc. For performance improvements, parallel and distributed databases have been designed [Bertino and Özsu, 1994], [Ozsu and Valduriez, 1999]. Still, data are coming from data sources (human, computer process or sensing devices) and stored in databases. Access is done via the databases interfaces. So why does the community in Grids speak about data sources again?

To our understanding, the problem comes from some of the constraints expressed before. Indeed these constraints of the Grids environments make it difficult to design a database management systems at a Grid scale, as you will see in the following the discussion, mainly for respecting the ACID (atomicity, consistency, isolation and durability) properties on the transactions in databases. Databases were not designed to handle at high throughput the amount of data found in Grids, especially when taking into account the security and multi-organization aspects of the access patterns. Moreover techniques from parallel databases to optimize joins in placing or moving data were less efficient at large scale when the network latency and bandwidth for instance were degraded. Also, it must be agreed that complicated access patterns are of less concern for the majority of the users of the Grids. These want to aggregate mainly computing and networking capacities, at the drawback of managing less efficiently the data. Hence for a long time, data concerns have been put behind the main track: In fact from 1996 to 2001, nearly no significant work has been done on the data management in Grids [Stockinger et al., 2001], [Chervenak et al., 2001], [Hoschek et al., 2000] at a level comparable to what was done in parallel computing some years before.

As we will see in Section 4.6, some initiatives try to push towards grid level databases, increasing autonomy and self management services, so that we can imagine again to forget about data sources.

4.3   Positioning the data management in grids within distributed systems

Data management in grids shares a number of aspects with some other systems dealing with data in distributed systems. In this section we will delineate data management in Grids by briefly comparing with Distributed Databases, Peer 2 Peer systems and Content Delivery Networks.

In distributed databases management systems (DDBMS) [Ozsu and Valduriez, 1999], storage devices are present on a set of nodes (potentially at the same site or at remote locations). The database is distributed into separate partitions, each one potentially replicated. The system makes sure that the distribution is transparent to the users and that the integrity of the transactions on the databases is ensured. Links between data pieces are explicit in the database schemas. This latter represents a big difference with the data management in grids where data pieces are generally more independent. A second difference concerns the security issues: Typically users in a DDBMS are handled by one organization or links between organizations have been set up once. The authorization checks take thus generally less time than in grids where ad hoc security services have to be contacted. A third difference is the network latency that can be high on grids, leading to poor performances for handling transactions. Even if this problem also appears in DDBMS in its principle, it is mainly considered that sufficient network connections exist and transactions can thus be held.

Peer 2 Peer systems (P2P) are widely developed and their main concerns have for a long time been to handle data sharing and access [Gribble et al., 2001]. Peers join and leave the infrastructure at any time, which is a key feature of P2P systems. Data management in P2P is thus mainly concerned about the search, the transport and availability of data. The completely decentralized approach of P2P systems is the core of the system: no global registry, no global resource management or data repository is existing. When peers cooperate for handling requests on data, they need to coordinate their representation of the data since a global schema can not be assumed. P2P systems can be seen as grids without central administration; but such a decentralized grid is far from being a reality mainly for global resources optimization and security concerns. Security and dependability of data in P2P systems are more problematic than in grids due to the higher dynamic of available resources and lack of central authority.

Content Delivery Networks (CDN) [Buyya et al., 2008] aim at providing an efficient access infrastructure for data pieces. They are mainly deployed on stable commercial infrastructure for multimedia data. They are developing techniques for the transport and the caching of data. Their main idea is to optimize the access of a subset of the whole data sets, predicting the ones that will be mostly accessed and organizing the replication and movement of data based on this fundamental. Quality of Services and realtime concerns are pinned by the multimedia nature of the data. Global indexes are generally maintained. Security has not been studied a lot in these dedicated environments. Data management in grids shares the main concerns of the CDN which, to some extent, can be seen as a subset of the data management part of the grids. Data management in grids is more complicated than the CDN. Indeed while the CDN is deployed on a stable infrastructure, the grids are more dynamic and the reliability of storage parties is limited. It can be noted that some CDN networks are deployed over P2P networks.

Finally a Grid storage can be seen hierarchically. There exists a local level, where the data are actually stored, either a file system running NFS, a database or a Storage Area Network (SAN). Second there is a global level where the data are transparently accessed, through primitives at the Grid middleware level. Several possibilities exist to create this high level transparent access (see Section 4.5): The main point is that there exist these two different access levels. Moreover, links between the different pieces of data are loose, and a global structure (or schema) is usually not existing. Finally, users accessing data come from different institutions and the transparent sharing of data in terms of security is a key feature in grids.

4.4   Links with the other services of the middleware

The Data Management in Grids has to be delineated in terms of functionalities. We will sketch some of these in the next section. It will by no means cover all the problems: For instance, the replica management procedures need some information about the possible sites where data can be stored.

Therefore, managing data in Grids relies on existing other pieces of a middleware that will provide some continuously updated input to the different services related to Data Management. The integration with other existing tools is of premier importance. These should be able to store and access their own functional data in a win-win schema.

Obvious services that interact with Data Management are those related with information and instrumentation of the state of the resources in a Grid. Monitoring and knowledge about the participating nodes, the available bandwidth, the latency, the available disk space (using grid level information systems [Czajkowski et al., 2001] or simple tools like the Network Weather Service [Wolski et al., 1999] or Network Distance Service [Gossa et al., 2007]) are necessary. Interaction with security services, from transport (SSL/TLS based), authentication (Globus Security Infrastructure [Foster and Kesselman, 2004] or simple LDAP), to authorization [Pearlman et al., 2003], [Seitz et al., 2005], [Alfieri et al., 2005]) are mandatory to secure the access to data, as we will detail later in Section 4.5.5.

Conversely, some services use data management procedures to perform their role: For instance the resource allocation manager (Globus Resource Allocation Manager [Foster and Kesselman, 2004], or OAR [Capit et al., 2005]) needs to retrieve information about the system state and may interact directly with the data managers.

These were only small and limited examples. Enumerating such links would be endless.

In brief, all the services of a running middleware benefit from and benefit to the services related to data management. This must therefore be carefully handled, with well defined and exposed interfaces and API.

4.5   Problems and some solutions

4.5.1   Data identification, indexing, metadata

The first problem that arises is a way to identify the data that will be managed and used in a Grid.[3] Typically a task on a Grid will need some input data and/or produce some output data. These data have to be stored, accessed, eventually moved from one site to another. There is thus a need to name the data, both individually and in group.

A task may request accesses to data stored at a distant site. Hence the naming must be global at the scale of a Grid in order to avoid mixing up different pieces of data. It must be coordinated so that different pieces of data get different global names. The naming should hide the different local naming conventions and organizations of the data storage: For instance, the underlying storage system may organize its storage hierarchically (trees for instance) or flat. The name of the data should not include information about their structures. Finally the naming must not be attached to a particular site as the data may move or may be replicated during their lifetime.

Using URN (Universal Resource Name, RFC 2141) to name the resources in Grids (and the data) is a solution more and more adopted in Grid middleware. Note that in the early EU DataGrid project [Stockinger et al., 2003], the Logical File Name (LFN) was adopted for similar aims. The URN fulfills the requirements of independence in the location of data and the organization of the actual storage. A possible way to construct a URN is by using a UUID (Universally Unique ID, RFC 4122).

The link between logical names and actual data (or resource in general) has to be ensured by an accessible service, an online database or ad hoc procedures. The Open Grid Forum proposes for instance the Resource Namespace Service for this mapping.

A name is useful mainly if one wants to retrieve some data. Two approaches can be envisioned:

- placement as a matter of the content: In this approach that is favored in some P2P networks, the data retrieval is only done thanks to the placement of the data itself: The identifiers of the data give a route in an overlay network to the data. For instance the Freenet P2P network

---

[3]The problem of identifying and naming is not related only to data but to resources in general: In fact, we will concentrate the view on data but most of the material is also valid for other kind of resources in Grids.

is using this technique, whereas the main grid middlewares do not use it.

- indexing the data in a registry: In this approach, some indexing and cataloging must be provided in order to help the actual access to the data. The difficulty is to address this registration in a distributed environment, managing some possible replication of data. Also to take into account is the fault tolerance of the service providing the registry so that the access to the data is not lost when one part of the catalog is missing. Redundancy and replication of the indexing services help for the fault tolerance and for balancing the load between several servers. Different methods exist in grid middleware for handling these catalogs, from centralized to distributed approaches. In the former, one catalog is maintained containing information about available data. In the latter, the catalog is either replicated at several sites or split between different servers, each one getting a part of a global catalog.

But a name is often not enough. Additional information is attached to the original data in many cases. This information is called the metadata of the data. We have to separate between two kinds of metadata: content metadata and management metadata.

- content metadata: These metadata are directly related to the content of the data. For instance a medical image may have the date of acquisition, the name of the doctor, the type of imager producing the data. Annotations made by data users during the data lifetime are also somehow attached to the data, like for instance a practitioner putting a comment on the observation of a medical scanner image. These additional information add semantic rich value to the raw data and are at least as valuable than the data themselves. Cataloging must take these into account as well since metadata provide an efficient way to search among the avalanche of data.

- management metadata: These metadata are linked with the data and used for the management of the data in the distributed systems. They include for instance the number and locations of potential replicas of the data (see 4.5.4), the usage of the data during their lifetime (for auditing, accounting, billing, traceability support). These metadata are used internally by the data management systems, by the other middleware services and by system administrators.

Representations of the metadata in the catalogs are typically put into database systems and/or expressed in XML (files or databases) or registries like LDAP (Lightweight Directory Access Protocol) directory trees. When used in Grids, the LDAP directories hold all the information for the management of the elements of the Grids (number of nodes, types of nodes, users, monitoring, ...) additionally to the data metadata.

When dealing with thousands of files, it is cumbersome (or impossible) to handle individually names and metadata. When sensors in a High Particle Physics experiment produce raw files continuously, it is convenient to associate all of them with that experiment and thus to group the produced data under a same collection. A collection allows a set of data to share common characteristics helping the search for particular results. Also on a data management point of view, a collection can then be handled like one set: If a part of the collection is moved (for instance a user makes a request for it) or removed (for instance if the data proved to be damaged), it is likely that the rest of the collection will have the same future allowing to anticipate on (and to optimize) the data management.

Metadata, semantics, grouping are means to undergo the way from raw data storage to information retrieval. Users want to access data with high level queries: Raw data is most of the time useless. For further reading, a good example of an efficient tool for indexing and retrieving in data collections is the MCAT catalog of the SRB (Storage Resource Broker) project [University of North Carolina, 2009].

The next section addresses the link between knowing where are the data and how to effectively access these data.

4.5.2   Data access, interoperability, query processing, transactions

As explained earlier, data sources in Grids are distributed among several sites and heterogeneous in several aspects. This heterogeneity appears in:

- storage systems: Each site of a Grid would hold its own storage system: The underlying storage systems can be any means to store data on permanent devices. The most encountered storage systems are raw file systems, most of the time based on some Unix-like file system (AFS, EFS). Another observed option in the High Performance Storage System [Teaff et al., 1995] which gives optimized access to files mainly in cluster when the file size is large. Databases are also seen in Grids, from open source implementation like MySQL or PostGres to commercial products like DB2 or Oracle-11G, the latter being advertised as a grid database, and mainly offers transparent access to different storage sites.

- data organization: Data access is also depending on the organization of the data, in terms of schemas, internal structure. Files structured or semi-structured will provide for instance means to extract more easily metadata facilitating their management.

- technologies: The tools to access the data depend on the storage system type and organization of the data. Local or Global API (Posix-like), specific protocols (like RFIO commands for HPSS), remote procedure calls (RPC), HTTP, web/grid services are all complementary means to actually get data pieces.

To handle this heterogeneity and distribution, a key concept in Grid computing is transparency. The idea is to provide a unique way to access the data, regardless of the actual storage systems, data organization or underlying access technologies. Data virtualization allows to access data virtually through a specific interface, mapped then to the actual data access patterns. This requires obviously appropriate naming and indexing of the data pieces (see previous section).

Most of the implementations rely on a limited set of operations representing the standard/common operations for all kinds of storage, that can be mapped onto these.

Concerning file access, the idea is to develop Posix-like functions (open, close, read, write, seek, stat,...) over any actual storage system to hide their heterogeneity. Grid wide access operations can be performed over distributed resources, with tools like object-based file system (XtreemFS [Hupfeld et al., 2008], VisageFS [Thizbolt et al., 2007], GridFS [Bhardwaj and Sinha, 2005] or even NFSv4), but due to consistency checks for ACID properties and latency delays, these solutions give low performances with scenarios when deployed at large scale on a grid.

Other problems include versioning and synchronization of replicas other than distributed environment. Replication will be detailed in Section 4.5.4.

In some cases, the data pieces can be split between different grid sites (like in XtreemFS), each of them holding a part of the data pieces. While being very efficient for increasing performances (for instance using parallel transfers, see Section 4.5.3), it creates more complexity for checksum computations, auditing, pinning and backups.

As we already mentioned in a previous section, some operations are held on a set of data pieces rather than on single data pieces. Bulk operations that are found (like in SRB) deal with registration, loading, deletion. The Hierarchical Data Format (HDFv5) gives a standard characterization of groups of files in a container.

The Data Access and Integration (OGSA-DAI) initiative [Antonioletti et al., 2005] in the OGSA architecture aims at providing an uniform access and integration to several kind of data sources, from relational databases, web databases, XML or even semantic enriched RDF files. The idea is to provide some mediator/wrapper between the actual data sources and the grid (thus the clients) that hides the heterogeneity of the access methods. Then the access itself is done via API calls, SQL queries or web browser (like OGSA-WebDB [Pahlevi and Kojima, 2004]). At the moment OGSA-DAI supports MySQL, IBM DB2, Microsoft SQL Server, Oracle, PostgreSQL and the eXist XML database.

Query processing is the process to actually access the pieces of data. As data are distributed, possibly replicated, there is the need to have some specific ways to perform the query processing. Intra-query parallelism in distributed query processing (DQP) deals with the access to the data when the data related to a single query are distributed among several sources. Inter-query parallelism appears when concurrent accesses to the same query service are provided. In order to provide an efficient query processing engine, both sides must be investigated and the query scheduler optimizes the query planning thanks to environment information. For instance, critical operations like table joins in databases must be done in the right order and on the right data sources or replicas. This optimization problem is not due to the distributed nature of the grid and appears also in traditional databases, but the problem is additionally stressed with the dynamic of the system and potential high latency of the network. The mainframe is to search for the potential data in the system, move the data to a processing engine and then perform an operation, with potential optimization in the order of the operation or the selected available operation service [Lynden et al., 2009]. Solutions based on mobile agents [Morvan and Hameurlain, 2009] aim at moving the query to the data rather than moving the data to the query engine.

Transactions are another important part in traditional databases. The idea is to group a set of operations on the data sources in an atomic operation. If one of the operation fails, the whole transaction is cancelled. In Grids, mainly two problems cause these transactions operations to be difficult: performance and checkpointing. In order to cancel a transaction not being able to finish, there is the need to make some checkpointing of the data sources before individual operations so that they can be restored in case of failure. Without a global coordinator of the transaction process, with network high latency and several thousands of clients, it becomes difficult to ensure the correctness or the freshness of the data, thus leading to inconsistency in the data. This problem is increased when data are replicated, since in that case all replicas must be in the same state at the end of the transaction. Works like GridTP [Qi et al., 2004], Turker [Türker et al., 2005], HVEM transaction [Jung and Yeom, 2008] show that some partial solutions exist, relaxing some constraints on consistency or using some existing hierarchical structure in the data set. See [Wang et al., 2008] for an historic survey of transaction management from flat to grid transactions.

Access is not enough: One should not forget how to transfer, how to control (security), how to process: GridFS for instance combines the benefits of different approaches (Remote File Access, File Staging) and offers special functionalities for Grid Computing (estimated transfer cost and file copy to local system).

4.5.3 Transport

While data transport is not directly related to data management itself, it is nevertheless useful to spend some time to describe briefly its links with data management. Besides the classical data access through sockets, RPC or RMI calls, web services invocation, FTP procedures, grids have seen the emerging of dedicated services to transport data between sites in the grid or to the final clients. The idea behind these services is to provide some additional possibilities in terms of performance or security thanks to the grid structure and middleware.

As an example, parallel streams can be used for transferring efficiently one piece of data. Indeed, as data are replicated in the system, it is obvious that opening different data streams between machines will decrease the transfer time. Experiences show that even when the data piece is on one single node, the time to deliver is shortened with parallel streams. Another example concerns the compression and the encryption of data during the transfer. GridFTP [Allcock et al., 2005] (at operating systems low level) or the Globus Reliable File Transfer service and gLite File Transfer Service (at higher abstraction level of service oriented grid architecture) [Globus, 2009b], [EGEE, 2009] are typical existing efficient possibilities of transferring data in the Grids. In RLS, the reliability of the transfer is resistant to failures of the network or sites. If one site can not complete a transfer, a replica is selected and the transfer continues.

4.5.4 Placement, replication, caching

The placement of data is a decision that impacts the performances of the Grids, both in terms of access time and robustness. In the DataGrid and now EGEE project like in many others, the solution was to adopt a hierarchical data structure. Subsets of data are copied at several levels of a data tree. The nodes of this tree are distributed to physical distant location: For instance in the LHC experiments of the DataGrid, CERN is the production center of the data. It is Tear 0. It is connected through high speed networks to several sites acting as Tier 1, having less capacities, where data are stored. These are recursively connected to Tier 2 with smaller bandwidth and data capacities.

It is common knowledge that replication of data increases performances, placing some data pieces closer to their future usage, decreasing the client waiting time and balancing the load between several potential servers. It also increases the reliability of the system, the data being more likely to be accessible even in case of system failures.

Failures that can show up are very diverse and account for several ways to replicate the data, which are easily possible in a heterogeneous grid. Media failure (disk failure, tape failure) can be overcome by replicating on multiple media. Avoiding problems with vendor specific system error, data should be replicated on different vendor products. To handle problems with the site connections, replication on a second site is necessary, while avoiding natural disasters requires the data to be replicated in a distant site. Finally, replicating data in deep archives (with more robustness and security) decreases the risks due to malicious users or noncareful administrators.

Works on replica management are numerous. They investigate several sides of the problem: from replica placement, migration, deletion, access to performances in terms of access time and disk capacity used. They try to balance the criteria of number of replicas, dynamic locations of replicas, local and global system performance .... Examples include the LCG Replica Catalog [gLite, 2009], the Globus Replica Location Service RLS and Data Replication Service [Globus, 2009a]. They allow registration and discovery of replicas using replica catalogs, mapping a logical identifier to actual physical locations of replicas. Choosing the best replica is a tedious task, based on acquired knowledge about the status of the infrastructure and needs of the users: Among others, the Network Distance Service [Gossa et al., 2007] helps to insert several contradictory parameters for data placement, data replication (number of replicas) or replicas selection. Being able to optimize the number of replica and their placement is a key issue, notably for keeping consistency manageable. Since several copies of the same data piece exist in the system, there is a need for consistency between these copies. This applies also with the metadata attached to the data. This aspect will be covered in Section 4.5.6.

Differences between replication and caching are subject to discussion. While replication is often made explicitly by a user or a service, caching operates the data pieces transparently for the system in order mainly to increase the performance of the access. Data in caches are stored temporarily and no guarantee on the presence of data pieces on some sites is given: The system can decide to delete arbitrarily the data to let space for another piece of data. Few works are interested in caching in Grids, since most claim that caching is just a specialized way for replicating. Nevertheless, the coordination of caches in a Grid like in [Cardenas et al., 2007] gives much benefit to all the sites of the grid especially where a community of users share interests and thus data pieces. It increases the caching capacities and allows for more advanced data management like for instance performance enhancement and data splitting between caches.

4.5.5   Security: transport, authentication, access control, encryption

Data security has to be ensured at different places in the architecture. The first thing is to ensure the security of the sites where the data are stored, then to ensure that the communication of the data is secure. In Grids, no specific work has been done concerning the secure transport of the data. The applications rely on well known protocols like SSL/TLS.

To secure sites, there is basically three complementary mechanisms: Authentication, authorization and encryption. The security problem is particularly difficult in Grids because of the lack of a global and centralized authority. Indeed, local administrators in autonomous sites of different organizations do not want to let the access decisions outside their control.

Authentication is the process of identifying securely the services and users that want to access the sites. It is not related to data management but is a necessary brick of a grid middleware where the data management services have to interact with distant sites on behalf of the users. To avoid the need for the user to authenticate manually on the different sites, single-sign-on procedures have been developed. They are mainly based on certification and delegation of authority, for instance by the use of proxy certificate like in the Globus Security Infrastructure GSI [Foster and Kesselman, 2004]. Shibboleth [Shibboleth, 2009] allows for the cooperation of identity servers and for identity federation. Traditionally the authentication of the users connecting to a service is done at the service site. In Shibboleth it is the user organization that verifies its presence in its database and transmits to the service the required attributes. GridShib [Scavo and Welch, 2007] is a grid version integrated in Globus.

Authorization allows for verifying the permissions to access specific resources (for instance data) when being authenticated on a site. Several access control mechanisms have been developed for Grids. The Community Authorization Service (CAS) [Pearlman et al., 2003] has been proposed in the Globus middleware to control the access to resources in a Virtual Organization (VO) in grids. Attribute certificates carry the users' membership in terms of VO. The organizations' members of a VO delegate the access control of some of their resources to the CAS servers. The CAS servers verify the certificate memberships and the resources authorities. Interconnecting VOs (by collaborative CASs) allows for statically mapping the different profiles implemented in each VO.

In VOMS (Virtual Organization Membership Service) [Alfieri et al., 2005] the authorization rules for accessing the resources stay at the resources sites: Thus the owner of the resource is responsible for its access control and not the VO administrator (opposite of CAS).

Permis (PErmission and Role Management Infrastructure Standards) [Chadwick et al., 2008] uses the Role Based Access Control (RBAC) to issue attribute certificate based on the role in an organization rather than to individuals. It integrates the delegation of authority where the SOA (Source of Authority) of the resource expresses the trusted entities allowed to issue attribute certificates.

In Sygn [Seitz et al., 2005], all permissions to access resources are encoded in attribute certificates that are stored with their owner. Sygn does not involve any communication when granting the access to the resources. At any time, resources administrators (or entities being delegated the permission) can issue and give new authorization certificates.

These approaches differ mainly at the granularity level of the access control, the location of the decision point of access control (centralized, replicated, distributed) and the responsibility of the resource administrator. They all use attribute certificates (X509 or ad-hoc).

Due to the lack of central authority, the access policies are normally not expressed in the same language. This heterogeneity has to be handled, for instance by the means of a standardized language to express the policies like

XACML (eXtensible Access Control Markup Language). Nevertheless, there is the need to work at a semantic level to potentially map the different policies.

The data access needs to be traceable. Indeed in many applications it is useful or mandatory to know the previous read-write access to the data.

Encryption secures the content of the data. It is recognized that access control does not protect against all the security threats. For instance a disk containing data might be directly accessed without using any middleware at maintenance sessions or when the site containing the data leaves the grid. Traditional encryption mechanisms can be used like RSA or DES, but then the secure management of encryption keys becomes difficult. Where should these be kept in order to allow access to authorized users or services while not suffering from single point of failures? Some works [Seitz et al., 2003] tend to distribute some part of decryption keys among a set of servers in order to decrease the risks related to an attacked key servers' repository.

Metadata are often attached to data. The security handling of both must be consistent, and this is also the case when data are replicated.

4.5.6 Consistency

The consistency problem appears when data are replicated and accessed within write patterns. Most of the huge amounts of data produced in production grids are mainly read-only. For instance data produced by the LHC experiments result from the collision of particles and thus are not subject from changes. From these data, additional data can be produced or computed, but usually their number is far smaller. As a result, data consistency did not gain much attraction in the past in the grid community.

The part where consistency made sense is the management of metadata. Some metadata are fixed like the acquisition date or the description of the experiment that produced the data in terms of instruments for instance. Some metadata may evolve with time, such as annotations made by experts after analysis of the data. Security permissions must also be considered as metadata and consistency should hold: When the permissions to access a data piece is modified, this modification has to be forwarded.

Since the support for consistency has long been mainly ignored in grid middleware, the consistency of the data was managed by the users themselves: These were replacing the modified data (or metadata) manually. This procedure is highly inefficient:

- The users can make some errors in the process, forgetting some replicas for instance;

- When several users wants to update data at the same time, they have to cooperate, leading to difficult user oriented coordination procedure.

- It is not robust when some sites are temporarily unavailable: These sites miss the updates of data leading to inconsistencies in the system.

Quorum mechanisms depending on the application requirements could be used to ensure that an operation is performed when at least a given number of replicas are available. Synchronization is then performed as soon as the replicas become again available.

Few works can be cited for the consistency management in grids and they clearly depend on the applications' requirements. [Pucciani, 2008] presents a comprehensive introduction, related works and innovative solutions. Solutions mostly rely on the concept of master replicas (up-to-date data are found at a master replica), and different mechanisms to take the modification of the master replicas (possibly several) into account [Sun and Xu, 2004], [Chang and Chang, 2006]: Lazy updates are done only when a slave replica (nonmaster replicas are slaves) is accessed; Push methods allow for a more aggressive way of updating data, at the initiative of the master replicas. Other methods [Chen et al., 2007] for databases are based on the possibilities of underlying replication strategies and consistency management in the databases, using the possibility to replay a set of operations in databases (with import and export logs).

4.6   Toward pervasive, autonomic and on-demand data management

Future works in data management in Grids include the integration of new concepts linked with the mobility, pervasiveness, context of the users and resources. Taking advantage of light devices, interconnected in an ad hoc way and participating with more stable resources to a grid is the key idea of the Pervasive Grid concept. The Pervasive Grid [Parashar and Pierson, 2009] encompasses many new challenges due mainly to the uncertainty of the information and resources on the next generation of grids. Early works on data management in these pervasive grids [Pierson, 2008] show the common approaches and differences between classical data management in grids and the data management in pervasive grids. The need for enhanced fault tolerance and recovery mechanisms, together with the inclusion of self characteristics (self-healing, self-management, self-recovery . . .) will lead to the development of a new class of grid computing, closer to autonomic computing [IBM, 2009]. Independent collaborative services (embedded in Service Oriented Architecture) will be developed and interconnected. Dynamic reconfiguration of components according to the evolution of context (like moving, replicating, splitting) will be more and more present in future developments.

Existing research directions exist for integrating data resources and computation resources. Indeed data are normally not used directly, but are processed before being delivered to users. Moving the data to the computation nodes can be inefficient compared to moving the processing to the data (like the mobile agent approach for Distributed Query Processing). OGSA-DAI opened the path for this integration (by defining operations during the execution of some requests). Subsequent works on data- and work-flow in grids [Glatard et al., 2005] paved the way for more efficiency and optimization. The idea is to allocate computation resources taking into account the placement of the data manipulated by the processes.

Another direction for data management in grids will be developed with the concept of Cloud Computing. Cloud computing accounts for integrating on-demand resources (Amazon for instance), up to deploying specific middleware (like Grid'5000 [Cappello et al., 2005]) on-the-fly for customers. Today such infrastructures are mainly based on dedicated clusters, but soon Grid Cloud Computing will become a reality. Data management in such environments will be based on works in grid computing, adding more efforts on quality of services and accounting for enabling comprehensive business models.

4.7 Concluding remarks

This chapter gave an overview of the different techniques related to data management in today's computing grids. It tried to sketch the fundamental differences with the data management in other distributed systems or to delineate the links with distributed databases and all the corresponding background. Exploring several problems and partly some solutions, we proposed a comprehensive view of the data management techniques for classical problems: Identification, Replication, Access, Query, Security, Consistency.

Finally, we gave briefly some future directions of the data management in grids towards autonomic and on-demand computing.

Acknowledgment

The author would like to thank his esteemed colleagues Lionel Brunie, Georges Da Costa, Abdelkader Hameurlain and Harald Kosch, and all those who, especially during the Data Management in Grids workshops [Pierson, 2005], [Pierson and Brunie, 2007], [Pierson and Kosch, 2008], raised some interesting and debated discussions in the last years.

4.8 References

[Alfieri et al., 2005] Alfieri, R., Cecchini, R., Ciaschini, V., dell'Agnello, L., Frohner, A., Lorentey, K., and Spataro, F. (2005). From gridmap-file to VOMS: managing authorization in a grid environment. *Future Generation Computer Systems*, 21(4):549–558.

[Allcock et al., 2005] Allcock, W., Bresnahan, J., Kettimuthu, R., and Link, M. (2005). The globus striped GridFTP framework and server. In *Proceedings of the 2005 ACM/IEEE Conference on Supercomputing (SC'05)*, page 54, Washington, DC, USA. IEEE Computer Society.

[Antonioletti et al., 2005] Antonioletti, M., Atkinson, M. P., Baxter, R. M., Borley, A., Hong, N. P. C., Collins, B., Hardman, N., Hume, A. C., Knox, A., Jackson, M., Krause, A., Laws, S., Magowan, J., Paton, N. W., Pearson, D., Sugden, T., Watson, P., and Westhead, M. (2005). The design and implementation of grid database services in OGSA-DAI. *Concurrency: Practice and Experience*, 17(2–4):357–376.

[Bertino and Özsu, 1994] Bertino, E. and Özsu, M. T. (1994). Guest editors' introduction. *Distributed and Parallel Databases*, 2(1):5–6.

[Bhardwaj and Sinha, 2005] Bhardwaj, D. and Sinha, M. (2005). GridFS: ensuring high-speed data transfer using massively parallel I/O. In Bhalla, S., editor, *Proceedings of the 4th International Workshop on Databases in Networked Information Systems (DNIS 2005)*, volume 3433 of *Lecture Notes in Computer Sciences*, pages 280–287. Springer-Verlag. Available online at: `http://springerlink.metapress.com/openurl.asp?genre=article{\&}issn=0302-9743{\&}volume=3433{\&}spage=280` (accessed May 1, 2009).

[Buyya et al., 2008] Buyya, R., Pathan, M., and Vakali, A., editors (2008). *Content delivery networks*. Springer-Verlag.

[Capit et al., 2005] Capit, N., Costa, G. D., Georgiou, Y., Huard, G., Martin, C., Mounié, G., Neyron, P., and Richard, O. (2005). A batch scheduler with high level components. In *Proceedings of the 5th International Symposium on Cluster Computing and the Grid (CCGrid 2005)*, pages 776–783. IEEE Computer Society.

[Cappello et al., 2005] Cappello, F., Caron, E., Daydé, M. J., Desprez, F., Jégou, Y., Primet, P. V.-B., Jeannot, E., Lanteri, S., Leduc, J., Melab, N., Mornet, G., Namyst, R., Quétier, B., and Richard, O. (2005). Grid'5000: a large scale and highly reconfigurable grid experimental testbed. In *Proceedings of the 6th IEEE/ACM International Conference on Grid Computing*

(GRID'2005), pages 99–106, Seattle, Washington, USA. IEEE Computer Society.

[Cardenas et al., 2007] Cardenas, Y., Pierson, J.-M., and Brunie, L. (2007). Management of a cooperative cache in grids with grid cache services. *Concurrency: Practice and Experience*, 19(16):2141–2155.

[Chadwick et al., 2008] Chadwick, D. W., Zhao, G., Otenko, S., Laborde, R., Su, L., and Nguyen, T.-A. (2008). PERMIS: a modular authorization infrastructure. *Concurrency: Practice and Experience*, 20(11):1341–1357.

[Chang and Chang, 2006] Chang, R.-S. and Chang, J.-S. (2006). Adaptable replica consistency service for data grids. In *Proceedings of the 3rd International Conference on Information Technology (ITNG'06)*, pages 646–651, Washington, DC, USA. IEEE Computer Society.

[Chen et al., 2007] Chen, Y., Berry, D., and Dantressangle, P. (2007). Transaction-based grid database replication. In *UK e-Science Al one Hands Meeting 2007*, Nottingham, UK.

[Chervenak et al., 2001] Chervenak, A., Foster, I., Kesselman, C., Salisbury, C., and Tuecke, S. (2001). The data grid: towards an architecture for the distributed management and analysis of large scientific datasets. *Journal of Network and Computer Applications*, 23:187–200.

[Czajkowski et al., 2001] Czajkowski, K., Kesselman, C., Fitzgerald, S., and Foster, I. T. (2001). Grid information services for distributed resource sharing. In *Proceedings of the 10th International Symposium on High Performance Distributed Computing (HPDC'2001)*, pages 181–194, San Francisco, USA. IEEE Computer Society. Available online at: `http://csdl.computer.org/comp/proceedings/hpdc/2001/1296/00/12960181abs.htm` (accessed May 1, 2009).

[EGEE, 2009] EGEE (2009). File transfer service. Available online at: `http://egee-jra1-dm.web.cern.ch/egee-jra1-dm/FTS/` (accessed May 1, 2009).

[Foster and Kesselman, 2004] Foster, I. and Kesselman, C., editors (2004). *The grid: blueprint for a new computing infrastructure*. Morgan Kaufmann, 2nd edition.

[Foster et al., 2001] Foster, I., Kesselman, C., and Tuecke, S. (2001). The anatomy of the grid: enabling scalable virtual organizations. *International Journal High Performance Supercomputer Applications*, 15(3):200–222.

[Glatard et al., 2005] Glatard, T., Montagnat, J., and Pennec, X. (2005). Grid-enabled workflows for data intensive medical applications. In *Proceedings of the 18th International Symposium on Computer-Based Medical Systems (ISCBMS)*, pages 537–542. IEEE Computer Society.

[gLite, 2009] gLite (2009). Documentation. Available online at: `http://glite.web.cern.ch/glite/` (accessed May 1, 2009).

[Globus, 2009a] Globus (2009a). Documentation. Available online at: `http://www.globus.org` (accessed May 1, 2009).

[Globus, 2009b] Globus (2009b). Reliable file transfer. Available online at: `http://www.globus.org/toolkit/docs/4.0/data/rft/` (accessed May 1, 2009).

[Gossa et al., 2007] Gossa, J., Pierson, J.-M., and Brunie, L. (2007). Adaptable distance-based decision-making support in dynamic cross-grid environment. In Kermarrec, A.-M., Bougé, L., and Priol, T., editors, *Proceedings of the 13th International EuroPar Conference (EuroPar'2007)*, volume 4641 of *Lecture Notes in Computer Sciences*, pages 437–446. Springer-Verlag.

[Gribble et al., 2001] Gribble, S. D., Halevy, A. Y., Ives, Z. G., Rodrig, M., and Suciu, D. (2001). What can database do for peer-to-peer ? In *WebDB*, pages 31–36.

[Hoschek et al., 2000] Hoschek, W., Jaén-Martínez, F. J., Samar, A., Stockinger, H., and Stockinger, K. (2000). Data management in an international data grid project. In *GRID*, pages 77–90. Available online at: `http://link.springer.de/link/service/series/0558/bibs/1971/19710077.htm` (accessed May 1, 2009).

[Hupfeld et al., 2008] Hupfeld, F., Cortes, T., Kolbeck, B., Stender, J., Focht, E., Hess, M., Malo, J., Marti, J., and Cesario, E. (2008). The XtreemFS architecture: a case for object-based file systems in grids. *Concurrency: Practice and Experience*, 20(17):2049–2060. Available online at: `http://dblp.uni-trier.de/db/journals/concurrency/concurrency20.html#HupfeldCKSFHMMC08` (accessed May 1, 2009).

[IBM, 2009] IBM (2009). Autonomic computing: IBM's perspective on the state of information technology. Available online at: `http://researchweb.watson.ibm.com/autonomic/` (accessed May 1, 2009).

[Jung and Yeom, 2008] Jung, I. Y. and Yeom, H. Y. (2008). An efficient and transparent transaction management based on the data workflow of HVEM data grid. In *Proceedings of the 6th International Workshop on Challenges of Large Applications in Distributed Environments (CLADE '08)*, pages 35–44, New York, NY, USA. ACM Press.

[Lynden et al., 2009] Lynden, S., Mukherjee, A., Hume, A. C., Fernandes, A. A. A., Paton, N. W., Sakellariou, R., and Watson, P. (2009). The design and implementation of OGSA-DQP: a service-based distributed query processor. *Future Generation Computer Systems*, 25(3):224–236.

[Morvan and Hameurlain, 2009] Morvan, F. and Hameurlain, A. (2009). Dynamic query optimization: towards decentralized methods. *International Journal of Intelligent Information and Database Systems*. Available online at: http://www.inderscience.com (accessed May 1, 2009).

[Ozsu and Valduriez, 1999] Ozsu, T. M. and Valduriez, P. (1999). *Principles of distributed database systems*. PrenticeHall, Englewood Cliffs, NJ, USA, 2nd edition. Available online at: http://www.amazon.ca/exec/obidos/redirect?tag=citeulike09-20\&path=ASIN/0136597076 (accessed May 1, 2009).

[Pahlevi and Kojima, 2004] Pahlevi, S. and Kojima, I. (2004). OGSA-WebDB: an OGSA-based system for bringing web databases into the grid. *Proceedings of the International Conference on Information Technology: Coding and Computing (ITCC 2004)*, 2:105–109.

[Parashar and Pierson, 2009] Parashar, M. and Pierson, J.-M. (2009). Pervasive grids: challenges and opportunities. In *Handbook of Research on Scalable Computing Technologies*. IGI Global.

[Pearlman et al., 2003] Pearlman, L., Kesselman, C., Welch, V., Foster, I., and Tuecke, S. (2003). The community authorization service: status and future. In *Proceedings of Computing in High Energy Physics (CHEP '03)*.

[Pierson, 2005] Pierson, J.-M., editor (2005). *Proceedings of the 1st Workshop on VLDB Data Management (VLDB DMG'2005)*, volume 3836 of *Lecture Notes in Computer Sciences*. Springer-Verlag.

[Pierson, 2008] Pierson, J.-M. (2008). Data management concerns in a pervasive grid. In *Proceedings of the International Conference on Vector and Parallel Processing (VECPAR)*, number 5336 in Lecture Notes in Computer Sciences, pages 506–520. Springer-Verlag. Available online at: http://www.springerlink.com (accessed May 1, 2009).

[Pierson and Brunie, 2007] Pierson, J.-M. and Brunie, L., editors (2007). *Proceedings of the Workshop on VLDB Data Management in Grids (VLDB DMG'2006)*, volume 19.

[Pierson and Kosch, 2008] Pierson, J.-M. and Kosch, H., editors (2008). *Proceedings of the Workshop on VLDB Data Management in Grids Workshop (VLDB DMG 2007)*, volume 20.

[Pucciani, 2008] Pucciani, G. (2008). *The replica consistency problem in data grids*. PhD thesis, University of Pisa, Pisa, Italy.

[Qi et al., 2004] Qi, Z., You, J., Jin, Y., and Tang, F. (2004). GridTP services for grid transaction processing. In Li, M., Sun, X.-H., Deng, Q., and Ni, J., editors, *Proceedings of the Second International Workshop* on Grid and Cooperative Computing (GCC 2003), volume 3033 of *Lecture Notes in Computer Sciences*, pages 891–894. Springer-Verlag. Available online at: `http://springerlink.metapress.com/openurl.asp?genre=article{\&}issn=0302-9743{\&}volume=3033{\&}spage=891` (accessed May 1, 2009).

[Scavo and Welch, 2007] Scavo, T. and Welch, V. (2007). A grid authorization model for science gateways. In *Proceedings of the Workshop on Grid Computing Environments (GCE)*.

[Seitz et al., 2003] Seitz, L., Pierson, J.-M., and Brunie, L. (2003). Key management for encrypted data storage in distributed systems. In *Proceedings of the 2nd International IEEE Security in Storage Workshop (SISW 2003)*, pages 20–30. IEEE Computer Society. Available online at: `http://csdl.computer.org/comp/proceedings/sisw/2003/2059/00/20590020abs.htm` (accessed May 1, 2009).

[Seitz et al., 2005] Seitz, L., Pierson, J.-M., and Brunie, L. (2005). Sygn: a certificate based access control in grid environments. Technical Report 2005-07, LIRIS.

[Shibboleth, 2009] Shibboleth (2009). Internet2. Available online at: `http://shibboleth.internet2.edu/` (accessed May 1, 2009).

[Stockinger et al., 2003] Stockinger, H., Donno, F., Laure, E., Muzaffar, S., Kunszt, P., and Millar, P. (2003). Grid data management in action: experience in running and supporting. In *Proceedings of the EU DataGrid Project on Computing in High Energy Physics (CHEP 2003)*, pages 24–28.

[Stockinger et al., 2001] Stockinger, H., Rana, O. F., Moore, R., and Merzky, A. (2001). Data management for grid environments. In Hertzberger, L. O., Hoekstra, A. G., and Williams, R., editors, *Proceedings of the 9th International Conference on High-Performance Computing and Networking (HPCN 2001)*, volume 2110 of *Lecture Notes in Computer Sciences*, pages 151–160. Springer-Verlag. Available online at: `http://link.springer.de/link/service/series/0558/bibs/2110/21100151.htm` (accessed May 1, 2009).

[Sun and Xu, 2004] Sun, Y. and Xu, Z. (2004). Grid replication coherence protocol. *Proceedings of the International Parallel and Distributed Processing Symposium*, 14:232.

[Teaff et al., 1995] Teaff, D., Watson, D., and Coyne, B. (1995). The architecture of the high performance storage system. In *Proceedings of the Goddard Conference on Mass Storage and Technologies*, pages 28–30.

[Thizbolt et al., 2007] Thizbolt, F., Ortiz, A., and M'zoughi, A. (2007). VisageFS: dynamic storage features for wide-area workflows. In Zheng, S., editor, *Proceedings of the International Conference on Parallel and Distributed Computing Systems (PDCS)*, pages 61–66. ACTA Press. Available online at: `http://www.actapress.com` (accessed May 1, 2009).

[Türker et al., 2005] Türker, C., Haller, K., Schuler, C., and Schek, H.-J. (2005). How can we support grid transactions ? Towards peer-to-peer transaction processing. In *CIDR*, pages 174–185. Available online at: `http://www.cidrdb.org/cidr2005/papers/P15.pdf` (accessed May 1, 2009).

[University of North Carolina, 2009] University of North Carolina (2009). Storage resource broker. Available online at: `http://www.sdsc.edu/srb` (accessed May 1, 2009).

[Wang et al., 2008] Wang, T., Vonk, J., Kratz, B., and Grefen, P. (2008). A survey on the history of transaction management: from flat to grid transactions. *Distributed Parallel Databases*, 23(3):235–270.

[Wolski et al., 1999] Wolski, R., Spring, N. T., and Hayes, J. (1999). The network weather service: a distributed resource performance forecasting service for metacomputing. *Future Generation Computer Systems*, 15(5–6):757–768.

Chapter 5

Future of grids resources management

Fei Teng

Applied Mathematics and Systems Laboratory, Ecole Centrale Paris, Grande Voie des Vignes, 92295 Châtenay-Malabry, France

Frédéric Magoulès

Applied Mathematics and Systems Laboratory, Ecole Centrale Paris, Grande Voie des Vignes, 92295 Châtenay-Malabry, France

5.1 Introduction

As network speeds grow, it is possible to construct large-scale high-performance distributed computing environments to allow users to submit jobs from anywhere in the world. These jobs can then be run on any available computing resource. As a consequence these resources should be assigned effectively to provide reliable and fast distributed services and to reduce the turn-around time of user jobs and the scheduling scheme should be heavily considered. There are numbers of heterogeneous or homogeneous clusters which can provide the job queuing mechanism, scheduling policy and local resource management, including utility computing, cluster computing, grid computing and so on. Among them, grid system is the most popular one and highly discussed, developed by the researchers and information technologies (IT) developers during the past decade. Recently a new term, "cloud computing," has emerged, which infers that computing is not operated on local computers, but on centralized facilities by third-party computing and storage utilities. Literally, clouds and other computing paradigm share the similar vision and object. They all aim to implement parallel computations that execute on many resources. However, clouds are more made available in a pay-as-you-go manner to the public or internal data center of business [Armbrust et al., 2009], [Buyya et al., 2000a], [Buyya et al., 2000b], [Buyya et al., 2001]. The cloud computing new characteristics differ from grid computing in the implementation details, which requires researchers and engineers to reconsider the resource scheduling strategies.

This chapter clarifies the differences of origin, aim and technology among several current computing paradigms including utility computing, grid computing, autonomic computing and cloud computing, focusing on their internal relationship, which implies the trend of development in the near future. The definition and architecture are then presented, which gives insights into the essential characters of cloud computing. Then, the most important issues are presented, i.e., service provision, introducing three different level services architecture and describing the attributes of cloud services. Upon them, resource management system appears to be the central component of cloud computing. How to build models which can exactly represents the clouds' current trends, such as resource-distributed and economic attributes, attracts our attention. After comparing cloud and grid system, we propose two new models for clouds: a resource-oriented model and an economy-oriented model. Finally, we try to predict some issues of resource scheduling where more efforts can be devoted in the near future.

5.2   Several computing paradigms

5.2.1   Utility computing

Utility computing was initialized in the 1960s, when John McCarthy coined the computer utility as:

> "If computers of the kind I have advocated become the computers of the future, then computing may someday be organized as a public utility just as the telephone system is a public utility ... The computer utility could become the basis of a new and important industry."

Generally, utility computing considers the computing and storage resources as a metered service, like water, electricity, gas and telephone utility [Yeo et al., 2006], [Paleologo, 2004], [Rappa, 2004]. The customers can use the utility services immediately whenever and wherever they need without paying for the initial cost of the devices. Utility computing is similar to virtualization so that the amount of storage or computing power available is considerably larger than that of a single time-sharing computer. The back-end servers such as computer cluster and supercomputer are used to realize the virtual-ization [Broberg et al., 2008]. From the late 90s, utility computing turns re-surfaced. HP launched the utility data center to provide the IP billing-on-tap services [HP, 2004]. PolyServe Inc. offers a clustered file system based on commodity server and storage hardware that creates highly available utility computing environments for mission-critical applications and workload opti-mized solutions specifically tuned for bulk storage, high-performance comput-ing, vertical industries such as financial services, seismic analysis and content serving. Thanks to these utilities, including database and file service, cus-tomers can independently add servers or storage as needed.

5.2.2 Grid computing

Grid computing emerged in the mid 90s. Ian Foster et al. integrated distributed computing, object-oriented programming and web services to coin the grid computing infrastructure [Foster and Kesselman, 2004], [Foster et al., 2002]. From then on, a lot of researchers gave the notion of grid computing in various ways. Here, we choose the definition of R. Buyya presented at the 2002 grid planet conference:

> "A grid is a type of parallel and distributed system that enables the
> sharing, selection, and aggregation of geographically distributed
> 'autonomous' resources dynamically at runtime depending on their
> availability, capability, performance, cost, and users' quality-of-
> service requirements."

This definition means that a grid is actually a cluster of networked, loosely coupled computers which works as a super and virtual mainframe to perform thousands of tasks. It also can divide the huge application job to several sub-jobs and make each run on large-scale machines. Generally speaking, grid computing goes through three different generations [Magoulès et al., 2008]. The first generation was marked by an early metacomputing envi-ronment, such as Fafner and I-Way. The second generation was represented by the development of core grid technologies, grid resource management—e.g., Globus, Legion—resource brokers and schedulers—e.g., Ccondor, PBS—and grid portals—e.g., Grid Sphere. The third generation saw the convergence between grid computing and web services technologies—e.g., WSRF, OGSI. It moved to a more service oriented approach that exposes the grid protocols using web service standards [Foster et al., 2001], [Shiers, 2009].

5.2.3 Autonomic computing

Autonomic computing was first proposed by IBM in 2001 with the following definition:

> "Autonomic computing performs tasks that IT professionals
> choose to delegate to the technology according to policies, see

[Liu et al., 2005].   Adaptable policy–rather than hard-coded
procedure–determines the types of decisions and actions that autonomic capabilities perform."

Concerning the sharp increasing number of devices, the heterogeneous, distributed computing systems are more and more difficult to anticipate, to design and to maintain the complexity of interactions. The complexity of management turns out to be a limiting factor of future development. Autonomic
computing focuses on the self-management ability of the computer system. It
will overcome the rapidly growing complexity of computing systems management and reduce the barrier that complexity poses to further growth.

In the area of multi-agent systems, several self-regulating frameworks are
proposed, but most of these architectures are centralized which mainly reduce management costs and seldom consider enabling complex software systems and providing innovative services [Jin and Liu, 2004]. IBM defined the
self-managing system, which can automatically process including configuration of the components (self-configuration), automatic monitoring and control
of the resources (self-healing), monitoring and optimizing the resources (self-
optimization) and proactive identification and protection from arbitrary at-
tacks (self-protection), only with the input information of policies defined by
humans. In other words, the autonomic system uses high-level rules to check
and optimize its status and automatically adapt itself to changing conditions.

5.2.4   Cloud computing

The cloud computing emerges as a new computing paradigm to provide
reliable, customized and quality of service guaranteed dynamic computing
environments for end-users [Weiss, 2007]. It is often confused with several
computing paradigms such as grid computing, utility computing and autonomic computing. According to the above description, we can draw the relationship among them. Utility computing cares that the packing computing
resources can be used as a metered service on the basis of the user's need. It is
independent of the organization of the resources, both in the centralized and
distributed system [Buyya et al., 2002]. But now, the companies prefer to
bundle the resources of members to provide utility computing. Grid computing is conceptually similar to the canonical definition of cloud computing, but
it doesn't manage the economic entities as well as it is less scalable than cloud
computing. Because of this massive scale, cloud computing must pay high attention on the interconnectivity management. In summary, cloud computing
depends on grids, has autonomic characteristics and utilities bills which can
be seen as a natural next step from the grid-utility model.

The computing paradigm varies with times. As shown in Figure 5.1, utility
computing was discussed frequently between 2004 and 2005. As a popular
term, grid computing is losing its appeal now. The term cloud emerged in
2007, and came to be the hot topic both in the research and industry domain.

From the day it was born, cloud computing overpassed grid computing and became more and more popular. Heaps of industry projects have been started including Amazon elastic compute cloud, IBM's blue cloud, and Microsoft's Windows Azure. At the same time, HP, Intel Corporation and Yahoo! Inc. recently announced the creation of a global, multi-data center devoted to open source cloud computing test bed for industry, research and education.

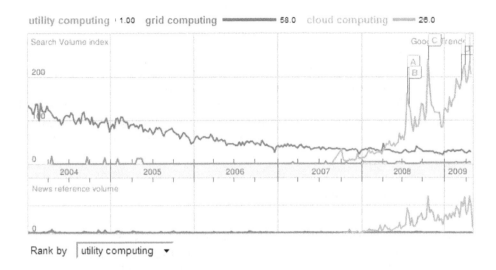

utility computing 1.00   grid computing ▬▬▬▬ 58.0   cloud computing ▬▬ 26.0

FIGURE 5.1: Google search trends for the last 5 years.

In order to analyze the reasons why cloud computing attracts so many researchers, we will firstly clarify the definition of cloud computing in the following section.

5.3   Definition of cloud computing

5.3.1   One definition

Since 2007, the term, "cloud" has become one of the buzz words in IT industry, while lots of researchers continue to define the cloud computing from different application aspects. Until now, there is no fixed consensus definition on it, and here we choose the definition of I. Foster [Foster et al., 2008]:

"A large-scale of e-distributed computing paradigm that is driven by economies of scale, in which a pool of abstracted virtualized, dynamically-scalable, managed computing power, storage, platforms, and services are delivered on demand to external customers over the Internet."

From the above definition, we can conclude the specific characters of cloud from grid computing. The resources in the grid system are located in the multiple administration domains, while the ownership in the cloud system is single. From interconnection aspect, the networks of the grids are always with high latency and low bandwidth. On the opposite, the clouds tend to be more dedicated to high-end with low latency and high bandwidth. What's more, the schedulers of grid systems focus on enhancing the performance of a specific application to meet its end-users' quality of service requirements. However, the scheduling scheme for clouds combine enhancing the performance of overall system and specific user. At the same time cloud computing strongly supports virtualization, dynamically compose services with web service interfaces, and so on.

5.3.2   Architecture

Clouds are usually referred to as a large pool of computing and/or storage resources, which can be accessed via standard protocols via an abstract interface. There is four-layer architecture for cloud computing as shown in Figure 5.2. The fabric layer contains the raw hardware level resources, such as computing resources, storage resources and network resources. The unified resource layer contains resources that have been virtualized so that they can be exposed to upper layer and end users as integrated resources. The platform layer adds on a collection of specialized tools, middleware and services on top of the unified resources to provide a development and/or deployment platform. The application layer contains the applications that would run in the clouds [Foster et al., 2008].

5.4   Cloud services

5.4.1   Three-level services

Generally, cloud computing can provide three-level services where the customers can choose one or more special service as they wish. For their special use, the customers can rent hardware, software or data as a service. Thereafter an integrated computing platform as a service is available. At the highest level, infrastructure used as a service is provided.

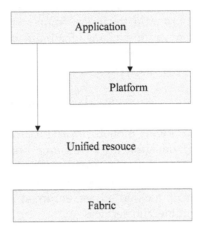

FIGURE 5.2: Cloud protocol architecture.

- Hardware as a service (HaaS) / Software as a service (SaaS) / Data as a service (DaaS):

- Hardware as a service (HaaS) is a pay-as-you-go model for accessing a provider's IT hardware, or even an entire data center. Some companies sell use of their hardware over the Internet on a per-use basis [Wang et al., 2008]. The user sends data and a program to process that data, while the vendor's computer does the processing and returns the result. The HaaS is flexible, scalable and manageable to meet the user's needs. An example is the IBM blue cloud project [IBM, 2007]. IBM delivers hardware infrastructure, database management, monitoring, security, availability and contingencies to be consumed by clients and partners directly.

- Software as a service (SaaS) is hosting software or application as a service and provided to customers across the Internet. This mode eliminates the need to install and run the application on the customer's local computers [Church et al., 2008]. SaaS therefore alleviates the customer's burden of software maintenance, and reduces the expense of software purchases by on-demand pricing. An example of the SaaS is Microsoft's software plus service, which combines local software and Internet services interacting with one another [Microsoft Corp., 2008].

- Data as a service (DaaS) is related to the fact that users can access remote data in various formats and from multiple sources. And then they operate them just like on a local disk. Typical example is Amazon simple storage service which provides a simple web services interface that can be used to store and retrieve, declared by Amazon, any amount of data, at any time, from anywhere on the web [Amazon Inc., 2008].

- Platform as a service (PaaS): Based on the support of the HaaS, SaaS and DaaS, the cloud computing in addition can deliver the platform as a Service for users. Platform as a service (PaaS) offers a high-level integrated environment to build, test and deploy custom applications. Generally, developers need to accept some restrictions on the type of software they can write in exchange for built-in application scalability. Google's App Engine enables users to build web applications on the same scalable systems that power Google applications [Google, 2008].

- IaaS: Infrastructure as a service (IaaS) provisions hardware, software and equipments to deliver software application environments with a resource usage-based pricing model. Infrastructure can scale up and down dynamically based on application resource needs. Amazon proposed elastic cloud computing service [Amazon Inc., 2009] which uses the Eucalyptus's open source cloud as the interface [Eucalyptus, 2009] to allow people to set up a cloud infrastructure at premise and experiment prior to buying commercial services.

Figure 5.3 shows the relationship between these services. Clouds provide services at three different levels. According to users' special demands, they can not only subscribe to their favorite computing services with requirements of hardware configuration, software installation and data access, but also be supplied with the higher level environment, integrated platform or open infrastructures.

5.4.2   Service characters

The services could be classified into several categories, as below:

- Dynamic provision: The provision of services is on-demand, which means that the users request the service by themselves near real-time without users having peak loads. Performance is monitored. Loosely-coupled architectures are constructed using web services as the system interface. At the same time, users can re-provision technological infrastructure resources rapidly and inexpensively.

- Economic pricing: As a kind of consumed utility, the capital expenditure is greatly reduced and converted to operational expenditure. The infrastructure is typically provided by a third-party and does not need to be purchased for one-time or infrequent intensive computing tasks, so the users have easier entry to the computing world. Pricing on a utility computing basis is fine-grained with usage-based options and no IT skills are required for implementation. Cloud providers should mask this pricing granularity with long-term, fixed price agreements considering the customer's convenience.

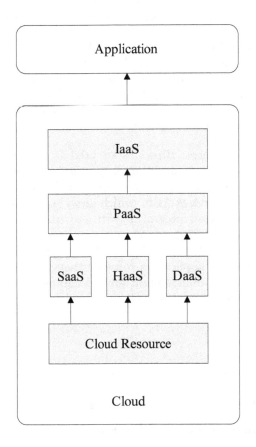

FIGURE 5.3: Three-level services.

- Security guarantee: Security typically improves due to centralization of data, increased security-focused resources, etc., but raises concerns about loss of control over certain sensitive data. Security is often as good as or better than traditional systems, in part because providers are able to devote resources to solve security issues that many customers cannot afford. Providers typically log accesses, but accessing the audit logs themselves might be difficult or impossible.

- Scalable location: In cloud computing, the users don't know the exact position where the services and devices are offered. The independent attribute makes users access systems using a web browser regardless of their location or what device they are using. The third-party infrastructure is accessed via the Internet; the users can connect from anywhere.

- Efficient utilization: The services in clouds can be compared as a large resource pool shared by users. The users not only can get lower costs because of the centralization of infrastructure, but also achieve higher peak-load capacity avoiding engineering for highest possible load-levels. What's more, through using the redundant sites, it's easy for business continuity and disaster recovery.

- User/system interfaces: Browsers are used as user interface which have the attributes such as intuitive, easy-to-use, standards-based, service-independent and multi-platform supported [Gens, 2008]. System interface adopts web services APIs, which provide a standards-based framework for accessing and integrating with and among cloud services. Cloud services provide well-defined, programmed access for users, partners and others who want to leverage the cloud service within a broader solution context. Thus, the application software that provides web-based GUIs, web services APIs, multi-tenant architecture and a rich variety of configuration options should be well developed in the future.

5.5   Cloud resource management

5.5.1   Comparison with grid systems

This section aims to compare the difference of resource management between grids and clouds. The first one is the business model for cloud system regards to the consumption! Like electricity, water or gas, the customer pays to the resource owner according to the amount consumed. On the contrary, grids have the project-oriented business model. The proposal represents the users who have a certain number of service units they can spend.

On the perspective of the computing model, the grids always use a local resource manager to manage the computing resources for a grid site, while the users submit jobs to request some resources for some time [Yu and Magoulès, 2009]. On the contrary, clouds share all the resources by all the users at the same time; that is to say, some low-latency applications can easily operate on clouds, which is not the case on grids.

Finally, the combination of the computing and data resource management is important. It is more efficient to schedule computational tasks close to the data, and to understand the costs of moving the work as opposed to moving the data [Chervenak et al., 2000]. Data-aware schedulers and dispersing data close to processors is critical in achieving good scalability and performance.

5.5.2 Resource model

The resource model is related to the question: "How to describe and manage resources in the system?" In the original resource model, the operation and data usually have two kinds of relations. One approach is related to data that comprise a resource and which are described in a specific description language along with some integrity constraints. One other approach treats operation on the resources as a part of the resource model. However in the cloud computing, because of the high virtualization of the resource and interacting operation of data, the operation and data must be considered as a whole part. Thus, a new resource model is proposed as represented in Figure 5.4.

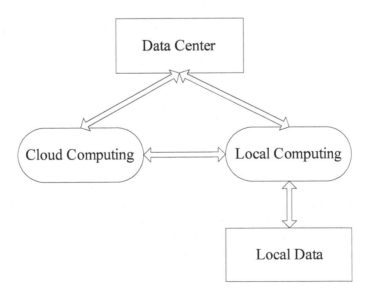

FIGURE 5.4: Resource model in clouds.

All the clouds users share resources to support the interactive applications. These applications are limited in grids because of the expensive scheduling decisions, data moving and potentially long queue times [Foster et al., 2008]. That is to say, computing will be centralized, when storage, operation and other kinds of resources are provisioned by clouds.

At the same time, local computing coexists with the cloud computing, and can be communicated and converted to each other if necessary. There are three reasons why we can't ignore the local operation. The first one is the Internet limitation. Even network technology is extremely developed today; some users can't access to the Internet anytime, or even won't suffer on the line all the time. So we must consider this situation in which the users prefer to finish their work when clouds are down. The second is related to the security consideration for some users, especially commercial enterprises, who don't want to run their security, critical tasks on the common clouds and send their sensitive data to the cloud storage. The third is related to some large-size companies well equipped with software and hardware, able to handle their data internally, which will be more effective than computing on clouds.

Virtual data in the center storage can be requested without regard to data location which considers data transparent to the users. The center data can either be computed in the clouds, or transferred to the special request. However, frequently staging data in and out to distant computers will slow down the computing speed. Besides, the speed difference of I/O and local disc to network storage can affect application performance. So we apply the local data which is most closed to the computing jobs as the solution to overcome the transmission problem. In summary, the resource model combines the data control and computing operation to minimize the amount of data movement and improve the end-application ability.

5.5.3   Economy-oriented model

Compared with other computing paradigm, cloud computing is more commercial and promises to deliver services on subscription-basis in a pay-as-you-go model. The traditional resource management always focuses on maximizing the throughput and minimizing the mean waiting time, but seldom includes important factors in the market such as the fair access to the resources [Yeo and Buyya, 2006], [Calheiros et al., 2009]. So we must orient economic models to enable on-demand trading of services and support customers buying the computing service like other utilities. The economy-oriented computing model is shown in Figure 5.5.

At the provider end, the lowest level implies numerous physical machines including all kinds of servers. The second layer contains virtual machines which utilize the physical machines to meet the customer's service request dramatically. Each virtual machine is isolated from each other on the same physical machine. Cloud services are abstracted as the actual applications operate on the highest level of the provider.

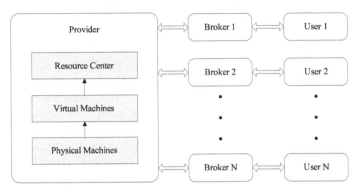

FIGURE 5.5: Economy-oriented model.

At the user-end, enterprise or individual users submit their service request and quality of service parameters from anywhere. It only waits for the reply of process from cloud provider, rather than directly deals with multiple heterogeneous providers.

Between provider and user, a broker bridges the two entities. The broker is equipped with a negotiation module that is informed by the current conditions of resource and current demands of user, so that it can help the users find a resource which meet its quality of service, and choose the user whose application can provide it maximum utility. So the broker will gain by the price difference between the user and provider. Furthermore, the broker will aggregate more jobs such as reserving resource slots, scheduling services and performing admission control to avoid overload.

5.6 Future direction of resource scheduling

Cloud computing is a long-held dream of computing as a utility, which has recently emerged as a commercial reality. Its system is larger, more heterogeneous and more dynamic than the original grids. Going along with the growth of cloud computing, there exist many obstacles which challenge the researchers and IT engineers to overcome them. Of course, each obstacle stands for an opportunity. Next, we will discuss the difficulties and trends for the resource scheduling.

5.6.1   Scalable and dynamic

Recently, grid or cloud systems contain more and more computing resources, storage services and application users and other distributed components which generate a massive scale environment in which the components can join or leave anytime as they wish. Besides, the condition of resources is temporal and the system can grow or shrink based on demand and operating environment. So both centralized and decentralized models for resource management should be enhanced at the same time. There is a need to develop scalable methods for resource discovery and scheduling that can adapt to changing resource and network conditions. On the other hand, past scheduling and resource allocation schemes are based on static approaches. Existing scheduling heuristics make the scheduling plan for the entire workflow in advance, and then the tasks are executed according to that. Considering the resource availability change and network condition variation, the methods that can perform as tasks arrive according to the current condition should be considered and developed.

5.6.2   Secure and trustable

Although grids and clouds share lots of technologies and philosophy, the biggest difference is a shift, from an infrastructure that delivers storage and computing resources to the one economy-based aiming to deliver more abstract services. Cloud computing is used in-house or open to the public. So it has high requirements in data integrity, recovery, privacy, regulatory compliance and auditing. Although virtual organizations could address some problems, for example, critical resource, personal and logistical issues, some individual customers or large enterprises still wouldn't like to run mission-critical applications on the cloud and send sensitive data to the cloud for processing and storage.

The infrastructure is autonomous, which means that every service domain makes its resource allocation decision independently and each one holds different aim and utility functions. In the current computing paradigm, trustworthiness and reliability of resources are seldom considered in the scheduling schemes. However, they are fundamental to ensure guaranteed service delivery. So we should pay more attention to scalable trust, reputation and security protocols in the resource scheduling algorithm.

5.6.3   Virtual machines-based

An obvious evolution from grid to cloud computing is virtualization which enables the applications to be isolated from the physical hardware. For example, a physical machine can be used as a set of multiple logical virtual machines, while the tasks can be run on any virtual machines. So on the same physical machine, we can host multiple operating system environments

separately and configure virtual machines to utilize different partitions of resources. Virtual machines-based technology gives challenges like the intelligent allocation of physical resources for managing competing resource demands of the users. Besides, the virtual machines and operating systems do
not provide a programmer-visible way to ensure all the application threads
to be run at the same time. Therefore, the future scheduling algorithm is
expected to address how to assign virtual machines to meet the changing
demand of resources by users as opposed to limited resources on a physical
machine.

5.7 Concluding remarks

In this chapter, we provided a short overview of current computing
paradigms, identifying each strength and weakness. Then, we evaluated the
key components of the cloud system to give readers a better understanding
about the dozens of different definitions of cloud computing. Services and
resources in cloud system have then been discussed. According to the new
requirements of cloud computing, layer-structured models are defined, which
can address the limitation of the existing models. However, all these ideal
models need further evaluation in our next step. In terms of the dynamic,
secure and VM-based features, which will bring challenges and opportunities
in clouds resource management domain, we discussed each of them in detail
and aim to make a contribution on the resource scheduling topic in the future.

5.8   References

[Amazon Inc., 2008] Amazon Inc. (2008). Amazon simple storage service. Available online at `http://aws.amazon.com/s3` (accessed May 1, 2009).

[Amazon Inc., 2009] Amazon Inc. (2009). Amazon elastic compute cloud. Available online at: `http://aws.amazon.com/ec2` (accessed May 1, 2009).

[Armbrust et al., 2009] Armbrust, M., Fox, A., Griffith, R., Joseph, A., Katz, R., Konwinski, A., Lee, G., Patterson, D., Rabkin, A., and Stoica, I. (2009). Above the clouds: A Berkeley view of cloud computing. Technical report, University of California, Berkeley.

[Broberg et al., 2008] Broberg, J., Venugopal, S., and Buyya, R. (2008). Market-oriented grids and utility computing: The state-of-the-art and future directions. *Journal of Grid Computing*, 6:255–276.

[Buyya et al., 2000a] Buyya, R., Abramson, D., and Giddy, J. (2000a). An economy driven resource management architecture for global computational power grids. In *Proceedings of the International Conference on Parallel and Distributed Processing Techniques and Applications (PDPTA 2000)*, Las Vegas, NV, USA.

[Buyya et al., 2002] Buyya, R., Abramson, D., Giddy, J., and Stockinger, H. (2002). Economic models for resource management and scheduling in grid computing. *Concurrency and computation: practice and experience*, 14:1507–1542.

[Buyya et al., 2000b] Buyya, R., Giddy, J., and Abramson, D. (2000b). *An evaluation of economy-based resource trading and scheduling on computational power grids for parameter sweep applications*. Kluwer International Series in Engineering and Computer Science.

[Buyya et al., 2001] Buyya, R., Stockinger, H., Giddy, J., and Abrams, D. (2001). Economic models for management of resources in grid computing. Technical report, Arxiv preprint cs/0106020.

[Calheiros et al., 2009] Calheiros, R., Ranjan, R., de Rose, C., Buyya, R., Trezentos, P., Yodaiken, V., Cabecinhas, F., and Lopes, N. (2009). Cloudsim: a novel framework for modeling and simulation of cloud computing infrastructures and services. Technical report, Arxiv preprint arXiv:0903.2525.

[Chervenak et al., 2000] Chervenak, A., Foster, I., Kesselman, C., Salisbury, C., and Tuecke, S. (2000). The data grid: Towards an architecture for the distributed management and analysis of large scientific datasets. *Journal of Network and Computer Applications*, 23:187–200.

[Church et al., 2008] Church, K., Hamilton, J., and Greenberg, A. (2008). On delivering embarassingly distributed cloud services. *ACM SIGCOMM Computer Communication Review.*

[Eucalyptus, 2009] Eucalyptus (2009). Eucalyptus project. Available online at: `http://open.eucalyptus.com/wiki/EucalyptusOverview` (accessed May 1, 2009).

[Foster and Kesselman, 2004] Foster, I. and Kesselman, C. (2004). *The grid: blueprint for a new computing infrastructure.* Morgan Kaufmann.

[Foster et al., 2002] Foster, I., Kesselman, C., Nick, J., and Tuecke, S. (2002). The physiology of the grid: an open grid services architecture for distributed systems integration. In *Proceedings of the Open Grid Service Infrastructure WG, Global Grid Forum, USA.*

[Foster et al., 2001] Foster, I., Kesselman, C., and Tuecke, S. (2001). The anatomy of the grid: enabling scalable virtual organizations. *International Journal of High Performance Computing Applications,* 15:200–222.

[Foster et al., 2008] Foster, I., Zhao, Y., Raicu, I., and Lu, S. (2008). Cloud computing and grid computing 360-degree compared. In *GCE'08: Proceedings of the Grid Computing Environments Workshop,* pages 1–10.

[Gens, 2008] Gens, F. (2008). Defining "cloud services" and "cloud computing". Available online at: `http://blogs.idc.com/ie/?p=190` (accessed May 1, 2009).

[Google, 2008] Google (2008). Google app engine. Available online at: `http://code.google.com/appengine/` (accessed May 1, 2009).

[HP, 2004] HP (2004). Hp utility data center. Available online at: `http://www.hp.com/#Product` (accessed May 1, 2009).

[IBM, 2007] IBM (2007). Ibm blue cloud project. Available online at: `http://www.ibm.com/developerworks/linux/library/l-cloud-computing/` (accessed May 1, 2009).

[Jin and Liu, 2004] Jin, X. and Liu, J. (2004). *From individual based modeling to autonomy oriented computation.* Lecture Notes in Computer Science.

[Liu et al., 2005] Liu, J., Jin, X., and Tsui, K. (2005). *Autonomy oriented computing: from problem solving to complex systems modeling.* Kluwer Academic Publishers.

[Magoulès et al., 2008] Magoulès, F., Nguyen, M., and Yu, L. (2008). *Grid resource management: Towards virtual and services compliant grid computing.* Chapman & Hall, CRC Press, Boca Raton, FL, USA.

[Microsoft Corp., 2008] Microsoft Corp. (2008). Microsoft software plus service. Available online at: `http://www.microsoft.com/serviceproviders/saas/default.mspx` (accessed May 1, 2009).

[Paleologo, 2004] Paleologo, G. (2004). Price-at-risk: A methodology for pricing utility computing services. *IBM Systems Journal*, 43:20–31.

[Rappa, 2004] Rappa, M. (2004). The utility business model and the future of computing services. *IBM Systems Journal*, 43:32–42.

[Shiers, 2009] Shiers, J. (2009). Grid today, clouds on the horizon. *Computer Physics Communications*, 180:559–563.

[Wang et al., 2008] Wang, L., Laszewski, G. V., Kunze, M., and Tao, J. (2008). Cloud computing: A perspective study. In *Proceedings of the 2008 Microsoft eScience Workshop*.

[Weiss, 2007] Weiss, A. (2007). Computing in the clouds. *ACM*, 11(4):16–25.

[Yeo and Buyya, 2006] Yeo, C. and Buyya, R. (2006). A taxonomy of market-based resource management systems for utility-driven cluster computing. *Software: Practice and Experience*, 36:1381–1419.

[Yeo et al., 2006] Yeo, C., de Assuncao, M., Yu, J., Sulistio, A., Venugopal, S., Placek, M., and Buyya, R. (2006). Utility computing and global grids. Technical report, Arxiv preprint cs/0605056.

[Yu and Magoulès, 2009] Yu, L. and Magoulès, F. (2009). Service scheduling and rescheduling in an applications integration framework. *Advances in Engineering Software*, 40:941–946.

Chapter 6

Fault-tolerance and availability awareness in computational grids

Xavier Besseron

INPG, 51 avenue Jean Kuntzmann, 38330 Montbonnot Saint Martin, 38041 Grenoble, Cedex 9 France

Mohamed-Slim Bouguerra

INRIA, 51 avenue Jean Kuntzmann, 38330 Montbonnot Saint Martin, 38041 Grenoble, Cedex 9 France

Thierry Gautier

INRIA, 51 avenue Jean Kuntzmann, 38330 Montbonnot Saint Martin, 38041 Grenoble, Cedex 9 France

Erik Saule

BioMedical Informatics, Ohio State University, 3190 Graves Hall, 333W 10th Avenue, Columbus OH 43210, USA

Denis Trystram

INPG, 51 avenue Jean Kuntzmann, 38330 Montbonnot Saint Martin, 38041 Grenoble, Cedex 9 France

6.1   Introduction

Machines we are using everyday are not perfect; they are often subject to dysfunctions. Such dysfunctions can have different sources such as processor's wear-out, mechanics part breaks in a hard drive, defective blocks in memory, cosmic rays altering the value of a bit in memory, etc. The purpose of this chapter is to investigate efficient solutions for guarantying the reliable execution of applications in computational grid platforms. We will put a special emphasis on the checkpointing techniques.

Since the last decade, computing systems turn to large scale platforms composed of thousands of processors. The characteristics of such new computing platforms evolved to more heterogeneity on both computing units and communication devices, to more unbalance between computing speeds and communication bandwidth, to hierarchical and versatile execution conditions. In short, they evolved to more complexity. At the same time, the applications followed this evolution, especially in the field of simulation of actual phenomena in Physics or Biology. The magnitude order of the running time of parallel applications is larger and larger, and they include more features that lead to more complex codes, for simulating phenomena always closer to reality. The consequence of this inescapable race toward complexity is that most actual parallel applications run on very large systems for long durations: It is not rare that they run several days or weeks.

As long as the execution of an application remains short and on few resources, the execution platform could be considered as static. Today, with the large scale computing platforms and the large actual applications, this assumption is no longer reasonable; dysfunctions can not be ignored or neglected. From official sources in IBM, when the BlueGene/L system with 65,536 nodes was designed, it was originally targeted to have one failure every ten days [Adiga, 2002]. Some time later, it was expected to have a mean time between failures less than 20 hours [Oliner et al., 2004] and more than ten failures a day occur in practice [Liang et al., 2006]. Nowadays, in similar large systems, one processor fails every hour. Thus, the developers or users of large applications expect several processors to fail during the execution of their applications. It is necessary to study, develop and analyze efficient strategies for providing a safe and reliable completion of large applications.

Fault-tolerant systems have been extensively studied in various domains. It has been historically studied in telecommunication or power networks and in critical embedded systems. In both cases, the goal is to provide a high availability of the service provided by the systems and is often achieved through redundancy by using multiple power providers, by ensuring that independent network paths exist or by checking computation with independent machines.

The basic concepts in safety [Avizienis et al., 2004] are the *fault*, the *error* and the *failure*. Let us introduce briefly these concepts. A fault is a dysfunction in a hardware or software component that leads to an error. A failure occurs when a system does not behave as it was expected. Failures are generated by errors, but the reader should be aware that an error does not always lead to a failure.[1] Usually, the failure of a component is the fault of a larger system. This cycle is standard in safety and a system that behaves correctly in presence of a processor failure is said to be *fault-tolerant* and the targeted property is to be failure-free.

To protect a system from failures, several approaches were proposed. Against byzantine faults, most of the techniques proposed are *ad hoc*. In the data or storage field, the code theory provides algorithms able to recover an error. With *algorithm-based fault tolerance* (ABFT), the fault tolerance scheme is tailored to the algorithm to be performed [Huang and Abraham, 1984]. This approach, under some assumptions, allows to correct errors due to faults in computations.

Faults that slow down an execution can be handled in several ways. It is for instance possible to handle them using the scheduling theory with uncertainty or with data perturbation. Robustness and sensibility analysis of scheduling algorithms can then provide reasonable solutions [Mahjoub et al., 2009]. There exist also some models where each task has a probability of being correctly executed. If a task fails, it must be executed again. In those models, it is frequent that several concurrent copies of a single task are allowed [Crutchfield et al., 2008].

Another way to deal with faults is to duplicate a given set of components. Several different ways of duplicating are used in function of whether each copy of each component is executed or not which are respectively called *active* and *passive* replication. In passive replication schemes, a classical technique is the *primary-backup approach* in which a copy of each component is provided but runs only if a failure has been detected. Active replication schemes schedule and execute several copies of each component. A failure detection mechanism is then used to choose which copies should be trusted [Assayad et al., 2004].

For the case we are interested in, a classical and efficient technique is to use *rollback-recovery* [Elnozahy et al., 2002]. The idea is to periodically make a snapshot of the distributed system and save it into a stable storage, i.e., which is supposed to be failure-free such as an external network attached storage system). When a failure is detected, the computation stops and restarts from the last checkpoint. Implementing this technique can be a difficult process since it involves complex algorithms that ensure that the stored global state is consistent [Chandy and Lamport, 1985].

The previous introduction stated the general context of failures in new computing platforms. The rest of the chapter is organized as follows. Section 6.2 starts with a discussion on the numerous aspects of faults, leading

---

[1] For instance, the result of a computation can be wrong. It can return a value of 2 instead of 1. But if the only interesting point is if the value is positive or negative, the error is invisible and does not generate a failure.

to different concepts. Then, we present the underlying model of computational grid and the main existing models of faults. The section ends by the definition of consistent states which is the basic notion in all fault tolerant mechanisms. Section 6.3 is devoted to the use of multi-objective scheduling as a way for optimizing the resources while guarantying a good reliability. The most popular approaches are analyzed (the main distinction is to allow or not the duplication of tasks for both permanent and transient faults). However, such approaches remain too dependent of the specificities of the underlying architectures; thus, usually, more generic mechanisms are preferred. Section 6.4 describes the protocols based on stable memory (i.e., log-based protocols and checkpointing). Such protocols can be modeled for obtaining expressions of the completion times (or expected completion times). We first present a survey of the most important models for the execution of an application in a parallel environment stressed by permanent failures without fault-tolerance mechanisms. Then, we show how to include the cost of stochastic checkpoints into the expression. Thus, using an adequate probability law of the arrival of faults, we obtain analytic expressions. These formulae allow to determine the optimal period between checkpoints and other related problems. Finally, we discuss more practical issues for implementing fault-tolerant protocols in actual parallel environments in Section 6.6. We propose a synthetic comparison of the main current implementations. We conclude this chapter by opening some challenging problems that should allow in the future to have more secure execution of parallel applications.

6.2   Background and definitions

In this section, a basic set of definitions is presented that will be used throughout the entire chapter. These definitions [Avizienis et al., 2004] give a characterization of the various concepts that come into play when addressing the dependability of grid systems. The basic qualitative definition of dependability is: "The ability to avoid service failures that are more frequent and more severe than it is acceptable to the users" [Avizienis et al., 2004]. More precisely, the dependability represents a set of attributes namely: Availability, Reliability, Robustness, Safety, Integrity and Maintainability. An exhaustive study of all dependability aspects of the system is therefore far beyond the scope of this chapter. This chapter focuses more on the availability which is the time proportion a system is in a functioning condition. More precisely, it is the probability that the system is in the correct state at a given time. We also focus on the reliability which represents the probability of failure in a given interval of time. Finally, the robustness corresponds to the ability of the system to behave as expected in the presence of failures. Thus, we will emphasize on the automatic or semi-automatic approaches to maximize these different attributes using fault tolerance techniques which are transparent to the application.

6.2.1 Grid architecture and execution model

The grid model abstracts the grid architecture, and then it allows to design and verify protocols. The grid system model (Figure 6.1) is a set of clusters which are interconnected through a wide area network (WAN). They are composed of individual computers interconnected together by a local area network (LAN). The clusters may have a network attached storage (NAS) connected to its LAN.

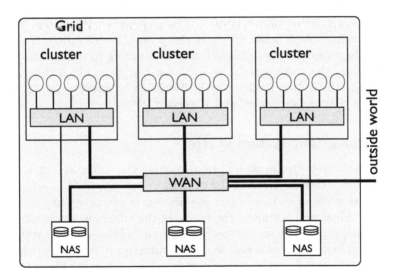

FIGURE 6.1: Grid system model: each individual node of a cluster is able to access to a network attached storage (NAS).

The distributed execution model consists in a set of processes that communicate only through messages (message passing model). The processes cooperate to solve a problem in a distributed fashion. They may interact with the outer world by sending or receiving messages. Some assumptions are required by the rollback-recovery protocols about the communication sub-system: most of the protocols assume that the delivery of messages is reliable and according to the FIFO order; some of them may accept message loss, duplication, or reorder.

6.2.2   Faults models

The failure is due to an *error* of the system which is a consequence of a *fault*. Different kinds of faults are usually distinguished in function of their origin and their temporal duration [Avizienis et al., 2004]. They could be intentional or not, software or hardware, modify the processing time of an operation, provide a wrong result or return no result at all. In this chapter, we are interested in accidental faults that do not modify the processing times of the computations and that provide no result in case of faults.

Faults can also be distinguished by the times during which they occur. A fault is said to be a *permanent fault* if the affected component will never behave correctly after the fault occurs or is said to be a *transient fault* if the fault is only active for a finite time interval. The length of the time interval of transient faults can be either deterministic or stochastic. It is frequent to consider transient faults of infinitely short durations.

Another important property of faults is the time when a given fault occurs. Despite existing cases where the fault arrivals are deterministic, it is more common to consider stochastic fault arrivals.[2] The concept of *mean time between failures* (often abbreviated MTBF) appears. When the faults are independent from each other or when the probability of failure is constant, it is usual to consider that the faults arrive according to a Poisson's process. In other cases, the more general Weibull law can be used. A description of several fault distribution models and a discussion on when to use them is the subject of a chapter in [Barlow and Proschan, 1996].

6.2.3   Consistent system states

Rollback-recovery protocols aim at restarting the execution after a failure from a global consistent state of the system.

The *global state* of a distributed application is composed of the states of all the individual processes and the states of the communication sub-system. State of each process could be easily captured.[3] However, the state of the communication subsystem is not accessible directly. It can be captured indirectly by flushing the communication channel or by logging the messages at emission or reception.

A *consistent system state* is a possible state of the system in a failure-free execution [Chandy and Lamport, 1985]. Applying the definition to the message passing model, a consistent system state means that "if a process state reflects a message receipt, then the state of the corresponding sender reflects sending that message" [Chandy and Lamport, 1985], [Elnozahy et al., 2002]. Let us consider for instance the global state C_2 of Figure 6.2. The

---

[2]A way to deal with deterministic faults is to use the scheduling theory in a model with machine availability.

[3]Multi-threaded processes may require more work.

process P_1 sends a message m_3 to process P_2. The global state C_2 is composed of the state of P_1 before sending and the state of P_2 after reception: it is inconsistent.

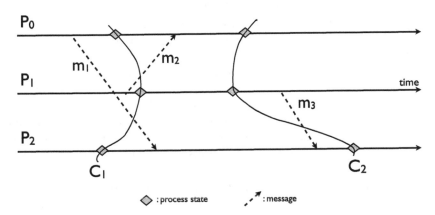

FIGURE 6.2: Three processes exchange messages. Two global states are considered: at the left the global state C_1 is consistent; at the right the global state C_2 is an inconsistent global state because message m_3 is received on process P_2 but not sent on process P_1.

6.3   Multi-objective scheduling for safety

6.3.1   Generalities

One common way to optimize the behavior of a parallel system is to optimize the resources by the use of the scheduling theory [Leung, 2004]. Basically, an application is represented as a set of tasks with constraints like precedences or communications. These tasks need to be distributed among a set of processors. There exist a lot of models for addressing the problem variants like processors with different speeds, specific routing topologies, etc. Some extra features can also be included into scheduling problems, like optimization of energy consumption or reliability.

Once the underlying model has been established, we can focus on the optimization of some performance indices on a target application, such as the makespan (defined as the maximum completion time of the application), or the flow time (the turnaround time in systems where data arrive infinitely). Within a scheduling formalism, it is possible to address the problem of safety as another objective, for instance, maximizing the reliability of the whole system. Some other works address the problem of optimizing of the number of tolerated faults. This issue will not be covered in this chapter but such kind of work can be found in [Benoit et al., 2008].

Optimizing only the reliability of a system does not really make sense since it will cost some processing time that will reduce the system performance. Then, the interesting problem is to optimize both the performance and the reliability of the system. In multi-objective optimization, the problem is to achieve a good trade-off between both objectives. This trade-off problem is solved by a decision maker and not by an automatic system. However, computers can be used to obtain an interesting set of solutions among which the decision maker will be able to choose the best one depending on his-her use. Here, interesting solutions belong to the Pareto set.[4] Details about multi-objective scheduling can be found in [Hoogeveen, 2004].

There are two common ways for determining the Pareto optimal solutions of a bi-objective optimization problem. The first one is to optimize a weighted sum of both objectives (or any other adequate combination). By changing the weight of both objectives, it should be possible to find the solutions that are on the convex hull of the Pareto set. The other way is to introduce a threshold on one objective that should not be exceeded, and then find the solution that is optimal for the second objective with standard single objective methods. The first method is generally easier from an algorithm perspective, but it will miss all the solutions that are not on the convex hull of the Pareto set. The second technique will provide all the Pareto optimal solutions but at the price of much more difficult optimization problems.

There are more pertinent models that can be detailed in this chapter. In the following sections, we focus on two scheduling models for safety in computational grids. Their applicability will be discussed. The problems they rise will be detailed as well as how to solve them. How to optimize safety as well as performance will be sketched. Since duplication leads to harder models, we will start by studying problems without duplication.

6.3.2   No duplication

We will present and discuss two problems depending on the type of faults. The first problem that we consider is to schedule an application with independent permanent faults distributed according to a Poisson's process.

---

[4]The Pareto set is the set of Pareto optimal solutions. Informally, a solution is Pareto optimal if no other solution is better on both objectives simultaneously.

Permanent faults is a common fault model in grid computing. Indeed, when a machine crashes, a technician will have to repair it. This manual intervention is usually longer than the computation itself. Faults distributed according to Poisson's process is a common assumption. It is like considering that the failure rate is constant over time. This assumption is not always realistic since usually the machines have an important failure rate when they are started for the first time (due to unstable hardware or configuration issues). Then, the failure rate will increase with hardware wear. However, if the computation is short compared to the grid lifespan, we can assume that the variability of the failure rate can be neglected. The last assumption is that faults are independent. This assumption is not realistic since several causes of fault are due to the environment such as network failure, power outage, air conditioning dysfunction, etc. However, this assumption is critical for computing the reliability of a system. The only way to prevent the effects of global dysfunctions is to backup the computations (for instance, using checkpointing).

The success probability of the application (i.e., the reliability) is the probability that all processors are still active when they complete their last task. This is a direct consequence of the permanent fault model. The optimization of both makespan and reliability of a parallel application on an heterogeneous platform (heterogeneous in computing power as well as in failure rate) has been considered [Dongarra et al., 2007], [Jeannot et al., 2008]. To optimize it, the concept of failure rate per operation is introduced. Optimizing the reliability is done by scheduling tasks on the most reliable (per operation) processor whereas optimizing the makespan is achieved by using the most powerful machines. The problem of scheduling independent task has been solved in [Jeannot et al., 2008] using an approximation algorithm that sets a threshold on the makespan. The idea is to consider the processors ordered by decreasing reliability and filling them up to the threshold. For the problem of scheduling an arbitrary application, [Dongarra et al., 2007] proposes a non-guaranteed heuristic using the same kind of techniques.

The second model that can be considered deals with independent transient faults distributed according to a Poisson's process. The idea behind transient fault is that the machine recovers. This model can not be directly applied to grid computing. It models the situation where another machine can be used to replace the one that crashed. Such a situation can happen in grids that are under-loaded or in computing systems where an operator guarantees the availability of the machine. Computing a reliability of a system in this model is somehow easier than before. It corresponds basically to the probability that each task is executed correctly. However, the network introduces some difficulties while computing the reliability. The problem is usually handled using heuristics with no approximation guaranty (greedy algorithms) that optimize a linear combination of both objectives [Dogan and Özguner, 2002].

6.3.3   Using duplication

Duplication makes things much more difficult to tackle but it also allows to improve drastically the reliability. The main existing model with duplication is based on an estimation of the reliability. Intuitively, without duplication, computing the reliability is done using very simple statistical events such as the processor is still active at the end of the schedule (for the permanent fault model), or during the lifespan of a task, the processor is active (for the transient fault model). With duplication, there is in general no simple events that allow to describe the success of an application execution. Like before, let us discuss the two main existing models.

Let us first consider the model of transient faults. One way for computing the reliability would be to consider all the possible combinations of fault occurrences and determine whether the schedule is still valid or not. This would consume a lot of computing power and, thus, it is not reasonable. A less time consuming way for computing the reliability is to consider that faults on tasks are statistically independent.[5] This assumption deduces that when a fault occurs, the impacted processor should recover before the next task begins. Then, it is possible to construct a causality dependence graph of an application. Each node of the graph represents the execution of one task on one processor and two additional nodes that represent the beginning and the end of the application are included. Directed edges represent the causality dependencies. If there is a path of tasks that are executed without fault from the beginning to the end, then the application is failure free. Such a graph is called a Reliability Block Diagram (RBD) [Lloyd and Lipow, 1962], [Siewiorek and Swarz, 1998].

Computing the reliability of the application can be done by computing the reliability of the RBD. Unfortunately, estimating the reliability of a RBD is, in general, an NP-Complete problem. However, if the RBD has some specific given structure, then the reliability can be computed in polynomial time. The main class of RBD with this property is the series-parallel graphs. The problem is then to construct a schedule so that the RBD is series-parallel. To reach this goal, it is necessary to add additional constraints to the scheduling problem.

A common property that meets this requirement is that when a task begins, all the copies of its predecessors have been taken into account. This breaks the hard part of combinatorics of the reliability estimation by making the RBD a series-parallel graph (in fact, a series of parallel macro blocks). This technique has been used twice depending on the communication pattern [Assayad et al., 2004 , Girault et al., 2009]. The structure of the reliability function usually makes approximation algorithm impossible to construct. However, efficient approximation algorithms were derived for some specific cases of independent tasks or for a chain of tasks in [Saule and Trystram, 2009].

---

[5]This assumption is different from the statistical independence of faults.

In a model with permanent faults, duplication is complex due to the strong statistical dependency of the faults on the tasks. It is likely that when a permanent fault appears, we would like to use a dynamic scheduling scheme that uses machines that are still functioning correctly. Obtaining strong results from the reliability point of view would broaden our comprehension of the problem. Otherwise, it is likely that efficient non guaranteed heuristics (within the theory of multi-objective scheduling) can be developed. Solutions based on work-stealing and checkpointing seem to be better suited to achieve efficient parallel executions even when the system is subject to faults.

To summarize, multi-objective scheduling can be applied to safety in order to help a decision maker to solve the trade-off between efficiency and reliability. This approach is general and needs to be adapted to different models. Unfortunately, not a single model is unanimously accepted since each computing system requires its own specific fault model. Each resulting problem is almost unique; tackling each of them would be a titanic work. Instead, theoretical approaches focus on general but reasonable assumptions that highlight the core properties practical heuristics should focus on.

6.4 Stable memory-based protocols

The most basic form of fault tolerance for parallel applications consists in rollback recovery using a stable memory. In order to ensure application reliability and to prevent computation loss, these protocols store the required information on the stable memory. In case of failure, this information allows recovering a coherent global state of the application at rollback. Rollback recovery techniques have received a great deal of attention from the research community.

In this section we focus on the presentation of the two categories of fault tolerance protocols based on a stable memory, which differ in the kind of information they store. The log-based protocols save non-deterministic events delivered to the processes, while the checkpoint-based protocols store the process states.

6.4.1 Log-based rollback recovery

Log-based protocols are based on the *PieceWise Deterministic* (PWD) assumption [Strom and Yemini, 1985]. PWD means that a process execution can be modeled as a sequence of state intervals; and that the execution during a state interval is deterministic. However, each state interval is initiated by a non-deterministic event.

Log-based protocols capture and log the non-deterministic events that initiated all state intervals. Then, when a process crashes, it can be recovered by (1) restoring it to the initial state and (2) replaying the logged events in the same order they appeared in the execution before the crash. To avoid a rollback to the initial state of a process and to limit the amount of non-deterministic events that need to be replayed, each process periodically saves its local state. Log-based mechanisms in which the only non-deterministic events in a system are the reception of messages are usually referred to as *message logging*.

Notice that logging protocols require a garbage collector to reclaim old logs that will never be used. The algorithm is more or less complex depending on the category of protocols. Thus logging protocols have interest to periodically save each process state in order to suppress old logs and speed-up recovery after a failure.

A disadvantage of log-based protocols for applications with extensive inter-process communication is the potential for large overhead with respect to space and time, due to the logging of messages. The most important advantage of log-based protocols is that they are well suited for a system that interacts with the outer world; these interactions are viewed as non-deterministic events.

Message logging protocols have to ensure that, once a crashed process has recovered, its state is consistent with the states of the other processes. This consistency property can be expressed as avoiding *orphan* processes. An orphan process is a process q that has received and delivered to an application a message from a crashed process p that has not logged the message as sent to q. Such a message is called an *orphan message*.

Logging protocols [Alvisi and Marzullo, 1998], [Elnozahy et al., 2002] can be classified as pessimistic, optimistic or causal depending on the way they prevent orphan processes.

6.4.1.1   Pessimistic logging

The principle of *pessimistic logging* protocols is to log any non-deterministic event on the stable omission before it affects the computation [Alvisi and Marzullo, 1998]. This prevents orphan processes and, then, reconstructing the state of a crashed process is straightforward as it only requires to replay these events.

Nevertheless, pessimistic protocols potentially block a process for each message it receives, even if no process ever crashes.

6.4.1.2   Optimistic logging

Other approaches that reduce the overhead of logging messages are the *optimistic logging* protocols. These protocols make the optimistic assumption that logging will be completed before a failure occurs [Strom and Yemini, 1985]. They kept a message log in a volatile log avoiding to wait as for pessimistic protocols. The volatile log is periodically flushed to the stable memory. Thus *optimistic* protocol does not enforce consistency of the process states during the execution if a failure occurs before the volatile message log is flushed. During recovery, a consistent state is computed by finding the last oldest state that eliminates all orphan messages.

6.4.1.3   Causal logging

Causal logging protocols [Elnozahy, 1993], [Alvisi and Marzullo, 1998] have the failure-free performance of optimistic logging while retaining most of the advantages of pessimistic logging. Each event can be logged asynchronously as in the optimistic protocol; and causal protocols still guaranty to recover from the last saved state.

For this purpose, they ensure that any non-deterministic event that causally precedes the state of a process is either stored in the stable memory or it is available locally for that process. The causality information is piggybacked with each message.

6.4.2   Checkpoint-based rollback recovery

Rather than logging events, checkpointing relies on periodically saving the state of the computation in stable storage [Chandy and Lamport, 1985]. If a fault occurs, the computation is restarted from one of the previously saved states. Thus, checkpointing-based methods differ in the way the processes are coordinated and in the derivation of a consistent global state. The consistent global state can be achieved either at the time of checkpointing or at the time of rollback recovery. The two approaches are called respectively coordinated and uncoordinated checkpointing.

6.4.2.1   Coordinated checkpointing

Coordinated checkpointing requires that all the processes coordinate the construction of a consistent global state before they write the individual checkpoints in a stable storage. The disadvantage is the latency and overhead associated with coordination. Its advantage is the simplified recovery without rollback propagation and minimal storage overhead, since each process only needs to keep the last checkpoint of the global "recovery line."

Coordinated checkpointing may be blocking [Tamir and Séquin, 1984] if the computation is stopped during the checkpoint phase or non-blocking such as the original Chandy Lamport protocol [Chandy and Lamport, 1985].

Blocking coordinated checkpointing is attractive because it simplifies the implementation. Even if blocking coordinated checkpointing protocol requires a global synchronization, the overhead of the protocol is mostly due to transfer data to the stable memory: in a short period of time, all processes checkpoint their states in the stable memory. In nowadays clusters, the size of the application state to checkpoint is very important due to the huge number of processors.

6.4.2.2   Uncoordinated checkpointing

With *uncoordinated checkpointing* protocols, each process independently saves its state and a consistent global state is achieved in the recovery phase [Elnozahy et al., 2002]. The advantage of this method is that each process can make a checkpoint when its state is small. However, there are two main disadvantages. First, there is a possibility of rollback propagation that cascades back to the beginning of the computation which is called the *domino effect* [Randell, 1975]. Second, each process may be required to store multiple checkpoints due to the cascading effect which makes the storage requirement much higher.

6.4.2.3   Communication-induced checkpointing

A compromise between coordinated and uncoordinated checkpointing is *communication-induced checkpointing*. To avoid the domino effect that can result from independent checkpoints of different processes, a consistent global state is achieved by forcing each process to take additional checkpoints based on some information piggybacked on the application messages [Baldoni, 1997]. The main disadvantage with this approach is that it results in the creation, and thus storage, of a large number of unused checkpoints, i.e., checkpoints that will never be used in the construction of a consistent global state.

All these fault tolerance protocols were implemented and experimented inside many middlewares, with MPI and other programming models. Existing implementations and comparisons are given in Section 6.6.

6.5   Stochastic checkpoint model analysis issues

Coordinated checkpointing is one of the most popular methods to improve the robustness of a computing system stressed by different kinds of faults and failures. Thus, a huge number of checkpoint/restart models appeared over the past decades to improve the performance of such a system and a lot of works in this field are still ongoing. In this section we survey the most important stochastic checkpointing models. We present the principle of algorithms and we discuss the main associated results.

We describe different models for the execution of an application without fault tolerance mechanisms in an environment stressed by permanent failures. These models use various hypotheses leading to analytical expressions like the expected completion time or like the variation of the completion time for the application. Then, we survey the most popular checkpoint/restart models and we show how fault tolerance techniques may improve the system performance.

6.5.1   Completion time without fault tolerance

In order to show how fault tolerance techniques improve the system robustness, we present some stochastic models that provide an analytical analysis of system performance. Such models provide the distribution of the completion time which leads to compute many important criteria (the expected completion time and the variation of the completion time) [Chimento and Trivedi, 1993], [Leung and Choo, 1984]. More precisely, such models are also useful for real-time systems where user requirements include deadlines on the completion time. In such a context, the computation of the expected completion time provides only limited information. Thus, we need to provide the distribution of the completion time since the expected completion time can not give the probability of tasks to miss the deadline.

To analyze these models, we consider the following notations and assumptions. The system is considered as a collection of CPUs connected by a network where user submits an initial amount of work denoted by ω. Under the failure-free assumption, the system takes T units of time to execute this amount of work. A classical execution is represented in Figure 6.3. Adding permanent failures assumption, after a period of time denoted by X_i the system goes down due to an unavailability state with a certain hazard rate denoted by $H_x(t)$. Then, R_i units of time are spent in debugging or repairing mode, before the system is available again. It is usual to suppose that this time is random and follows a hazard rate $H_r(t)$. In this analysis, fault detections are supposed to be without latency. If this is not the case, then it may still be possible to account for the failures detection delay by suitably increasing the availability time or the repair time distribution. To model both the time between failures X and the time needed to repair R, it is supposed that both of them are random variables that follow a distribution function denoted by $F_x(t)$ and $F_r(t)$ and a density function denoted by $f_x(t)$ and $f_r(t)$. Therefore, the usual relation between the hazard rate function and the distribution function is:

$$F_x(t) = 1 - e^{\int_0^t H_x(x)dx} \text{ and } F_r(t) = 1 - e^{\int_0^t H_r(x)dx}.$$

Under those different assumptions and abstractions, the global application execution scheme can be represented by a function giving the amount of work that still has to be done at each time. Such a function is represented in Figure 6.4. Then, the total elapsed time V until the completion of the application can be computed as follows:

$$V = X_1 + R_1 + X_2 + R_2 + \cdots + X_{n-1} + R_{n-1} + T. \tag{6.1}$$

Let us now consider particular distributions. For instance, suppose that the failure process is modeled by a Weibull law which is one of the most used laws to model failures [Schroeder and Gibson, 2006]. The density function of

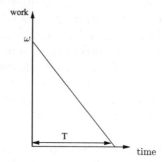

FIGURE 6.3: Execution scheme without failures.

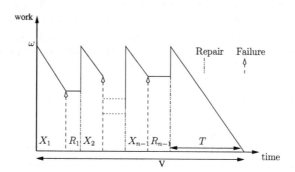

FIGURE 6.4: General execution scheme under failures and without fault-tolerance.

this law has two parameters λ, β and it is given by the following expression:

$$f_x(t) = \lambda\beta(\lambda t)^{\beta-1} e^{(-\lambda t)^\beta}.$$

• Let us first suppose that $\beta = 1$, i.e., the Weibull is reduced to an exponential law. Then, the expected completion time $\mathbb{E}(V)$ and the variance of the completion time $\mathbb{V}ar(V)$ have the following expressions:

$$\mathbb{E}(V) = (e^{\lambda T} - 1)(\mu_r + 1/\lambda). \tag{6.2}$$

$$\mathbb{V}ar(V) = (e^{\lambda T} - 1)[\mu_r^2(1 + \delta_r) + (e^{\lambda T} + 1)(\mu_r + 1/\lambda)^2] - \\ 2(\mu_r + 1/\lambda)[\mu_r(e^{\lambda T} - 1) + Te^{\lambda T})] \tag{6.3}$$

where $\mu_r = \mathbb{E}(R_1)$, $\delta_r = \sqrt{\mathbb{V}ar(R_1)}/\mu_r$ is the coefficient of variation of the restart time (the detailed analysis can be found in [Leung and Choo, 1984]).

• Secondly, let us suppose that $\beta > 1$ or $\beta < 1$, quadrature is generally required to compute the expected completion time or the variation of the completion time.

$$\mathbb{E}(V) = T + \frac{1}{1 - F(T)}[\int_0^T g f_x(g)dg + F_x(T)\mu_r] \tag{6.4}$$

Since the integral is always bounded by the first moment of time between failures X denoted by μ_x and the expected completion time must be at least T, we have:

$$T \leq \mathbb{E}(V) \leq T + (\mu_r + \mu_x)/(1 - F_x(T)). \tag{6.5}$$

Equation (6.2) states that the completion time of the application increases exponentially when ω grows. This confirms that fault tolerance techniques are required to improve the system performances. More generally, we conclude from Expression (6.5) that if most of the mass of f concentrates in the interval $[0, T]$ then $F_x(T) \to 1$, $\int_0^T g f_x(g)dg \to \mu_x$ and the completion time tends to be much longer than the completion time with the failures-free assumption. On the other hand, if most of the mass of f_x concentrates in the interval $[T, +\infty)$ then $F_x(T) \to 0$, $\int_0^T x f_x(x)dx \to 0$ and the completion time tends to be equal to T. Thus, fault tolerance techniques should be investigated in this context. Let us now introduce how the checkpoint/restart mechanism can improve the system performance.

6.5.2   Impact of checkpointing on the completion time

To deal with failures and to improve the system performance, many fault tolerance techniques have been proposed during the last decades. Based on simulation results, Elnozahy et al. [Elnozahy and Plank, 2004] showed that the coordinated checkpoint approach is the most effective fault tolerance mechanism in large parallel platforms. Let us recall briefly that the aim of such models is to optimize the trade-off between the amount of work lost when a failure occurs and the performance lost due to the checkpoint overhead with respect to some given metric. The most studied metric in computing systems is the completion time of an application. The fault-tolerance techniques are considered as a defensive mechanism added to the application to reduce the failures consequences (the amount of lost work due to failures in the case of checkpointing mechanisms). Unfortunately, these defensive mechanisms introduce also overheads that decrease the performance of the whole system. For instance, checkpointing on a BlueGene machine is reported to take an hour [Liang et al., 2006]. Hence, using these mechanisms without optimization techniques can decrease seriously the system performance. Since the 70s, several works model the application and checkpointing mechanisms and apply adapted analytical methods to optimize the system performance with respect to many criteria. All these models [Young, 1974], [Daly, 2006], [Ziv and Bruck, 1997], [Chandy and Ramamoorthy, 1972], [Toueg and Babaoglu, 1983] differ in some critical assumptions of the computing system.

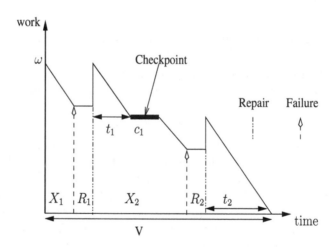

FIGURE 6.5: General scheme of an execution under failures with checkpoint mechanism.

The two following sections present an analysis of the checkpointing mechanism under two simple but reasonable hypothesis. The previous assumptions are still valid. In addition, we assume that the initial amount of work ω considered as a big task is preemptive (this is a mandatory assumption to implement checkpointing). The execution of this task will be divided into k consecutive intervals of length t_j such that $\sum_{j=1}^{k} t_j = T$. A checkpoint occurs between each interval and cost c_j units of time as it is depicted in Figure 6.5. The difference between both analyses is related to the cost function of a checkpoint. In the first case, it is supposed to be constant whereas it becomes variable in the second one.

6.5.2.1   Constant checkpoint cost

Young proposes in [Young, 1974] a checkpoint/restart model where failures follow a probabilistic law, but the checkpoint cost and the restart cost are constant. It also assumes that checkpoints can be placed at any moment of the application. Moreover, checkpointing and restart phases are assumed to be fault-free. Under those assumptions, Young provides a first order approximation of the optimal time between two successive checkpoints using the following arguments. Let O be the overhead due to the work during failures and the checkpoint cost, τ be the period between checkpoints such as $\tau = \frac{\omega}{k}$ and λ be the rate of the Poisson process which represents failures arrivals. Thus, O is given by the following expression:

$$O = 1/\lambda + \frac{c}{1 - e^{\lambda(c+\tau)}} \qquad (6.6)$$

Therefore, the optimal period between checkpoints τ is the root of the derivative function of Expression (6.6) which is equal to $\sqrt{2c/\lambda}$, considering a first approximation for the exponential term and assuming that the checkpoint cost is much shorter than the failures rate $c << 1/\lambda$.

Daly extends in [Daly, 2006] the model introduced by Young by proposing a higher order solution for the optimal interval of time between checkpoints considering that failures may happen during the checkpoint phase and the restart phase. In fact, he proposes another interval between checkpoints equivalent to Young's optimal interval length if the checkpoint cost less than $2/\lambda$; elsewhere the optimal interval length will be equal to the $1/\lambda$. Therefore, Daly's model is more precise than Young's model when the checkpoint cost is close to the mean time between failure $(1/\lambda)$.

$$\tau = \begin{cases} \sqrt{\frac{2c}{\lambda}}[1 + 1/3\sqrt{\frac{2c}{\lambda}} + 1/9(\frac{2c}{\lambda})] - c & \text{if } c \leq 2/\lambda \\ 1/\lambda & \text{if } c > 2/\lambda \end{cases} \qquad (6.7)$$

Considering both models shows that the average completion time will grow linearly when the initial amount of work grows. In fact, it is clear that the expression of the overhead due to failures and checkpoint according to Young's model [Young, 1974] does not depend on the initial amount of work which implies that the expected completion time will grow linearly. Also in [Daly, 2006] based on simulation the author reaches the same conclusion. Hence, checkpointing improves the system performance.

6.5.2.2   Variable checkpoint cost

The second variant of checkpoint/restart model are the models that consider a variable checkpoint cost. Several works claim that the checkpoint cost should not be considered as constant [Ziv and Bruck, 1997], [Chandy and Ramamoorthy, 1972], [Toueg and Babaoglu, 1983]. In fact, a popular technique to reduce the amount of data to save in the stable storage is to use an incremental method which only saves the memory which changed from the previous checkpoint. Using such a technique, considering a checkpoint cost is no longer a reasonable assumption.

The first work was proposed by Chandy et al. in 1972. Based on graph theory, it finds the optimal placement of the checkpoints [Chandy and Ramamoorthy, 1972]. The proposed technique relies on the existence of a prior-information about the checkpoint cost. Moreover, it also assumes that failures follow a Poisson's process.

Toueg et al. tackle the same problem under the following assumptions: the application can be only preempted at n specific times t_i for $1 < i < n$ and the failures follow a Poisson's process [Toueg and Babaoglu, 1983]. The cost of a checkpoint is then c_i. Under this model, they propose an $O(n^2)$ algorithm based on a dynamic programming that leads to an optimal expected completion time. The algorithm assumes that there are only n finite places to schedule the checkpoint. The recurrence objective function is based on Equation (6.8) that gives the expected completion time $\mathbb{E}(V)$ if a checkpoint is placed at the index i.

$$\mathbb{E}(V) = \frac{e^{\lambda(t_1+t_2+\cdots+t_i)} - 1}{\lambda} + c_i + \frac{e^{\lambda(t_{i+1}+\cdots+t_n)} - 1}{\lambda} \tag{6.8}$$

To find the optimal placement of different checkpoints, the algorithm iterates on the index i leading to the optimal sequence of checkpoints in n^2 iterations at most.

Another important contribution is proposed by Zvi et al. in [Ziv and Bruck, 1997] using the following assumptions: failures arrive following a Poisson's process and the application can be preempted at any time t to take a checkpoint. They assume that the system is modeled by a Markov chain composed of two different states s_1 and s_2 and of transition function ϕ (that is to say, ϕ_1 is the probability of going from state s_1 to state s_2 and ϕ_2 is the probability of going from state s_2 to state s_1). When the system is in state s_1 (resp. s_2), the checkpoint cost is c_1 (resp. s_2). They propose an algorithm to decide when a checkpoint should be taken. The algorithm has two parameters t_1 and t_2 such that $t_1 \leq t_2$ and is now stated:

repeat
    Wait t_1 units of time.
    if the state is s_1 **then**
        Take a checkpoint. The overhead is c_1.
    else if the state is s_2 **then**
        Wait up to t_2 for the system to change to state s_1.
        if the system changes to state s_1 **then**
            Take a checkpoint. The overhead is c_1.
        else
            Take a checkpoint. The overhead is c_2.
        end if
    end if
until all work is done

The average overhead ratio of this algorithm denoted by O is given by the following expression:

$$O = \frac{e^{\lambda t_1} + \frac{\lambda p_2}{\lambda - \phi_2}\left(e^{\lambda t_2} - e^{\lambda t_1 + \phi_2(t_2 - t_1)}\right) - 1}{\lambda\left(t_1 + \frac{p_2 e^{\phi_2(t_2 - t_1)} - 1}{\phi_2}\right)} + \frac{(1 - p_2)c_1 + p_2 c_2}{t_1 + \frac{p_2 e^{\phi_2(t_2 - t_1)} - 1}{\phi_2}} - 1 \quad (6.9)$$

where p_2 is the probability that the state at a checkpoint is s_2. The final step is to find the couple (t_1, t_2) that minimizes the average overhead. The proposed solution is to use a numerical method.

Let us summarize the results of this section. Under a basic stochastic model, the completion time of an application grows exponentially if no fault-tolerance techniques are used. Moreover, using some stochastic checkpoint/restart models, it is possible to improve the performance of orders of magnitude. However, these models are quite optimistic and many hypotheses do not hold in actual computing systems. In the next section we focus on the implementation issues of some fault tolerance mechanisms.

6.6 Implementations

A complete fault tolerant middleware is a complex system that consists in many interconnected components. This section presents first some implementations of single process checkpoints, and then an overview of implemented distributed fault tolerance protocols. Finally, we give a synthetic comparison of the main current implementations.

6.6.1 Single process snapshot

The problem of the single process checkpoint can be addressed following three approaches depending at which level this checkpoint is performed.

In the first approach, a process is saved as a memory dump. It can be done at the kernel-level with Berkeley Lab's Linux Checkpoint/Restart (BLCR) [Duell et al., 2002] or at the user-level with a library like Condor [Litzkow et al., 1997] or Libckpt [Plank et al., 1995]. Among these solutions, BLCR is the only one that supports multi-threaded applications. This method is widely employed since it is transparent for the application developer. But many drawbacks affect this kind of snapshot: it requires a homogeneous resource (same operating system and same CPU) when restarting. Furthermore, the process address space contains data that are not strictly required for restarting, so the checkpoint size is larger than necessary.

In order to abstract the process state, the second approach considers that it is the user responsibility to write functions for saving and restarting a process. This method is effective because the developer can choose exactly which data to checkpoint, but it requires a significant effort from the application developer.

The third approach proceeds at middleware level. It combines advantages from the two above approaches, but it requires the application to be written with a middleware that uses an abstract representation of the application, like objects (Charm++ [Zheng et al., 2004], [Chakravorty and Kalé, 2004]), task lists (Satin [Wrzesinska et al., 2006]) or data flow graphs (Kaapi [Besseron and Gautier, 2008], [Jafar et al., 2009]). Using the application abstract representation, the middleware can checkpoint on its own tasks and data used by the application. This approach is fully transparent for the application developer; the process can be restarted on a heterogeneous resource (since the abstract representation is architecture-independent) and the snapshot is smaller than in the memory space's case.

6.6.2 Fault-tolerance protocol implementations

The purpose of fault tolerance middlewares is to provide an easy way to make an application fault tolerant. Some approaches can be semi-automatic or fully automatic depending on the amount of work for the developer to adapt his-her application for fault tolerance.

Among the semi-automatic category, we find FT-MPI [Fagg et al., 2001] and LA-MPI [Aulwes et al., 2004] that tolerate communication failures. In case of failure, the error is reported at the application level to be treated and to react to the failure. Semi-automatic middlewares can offer good performances because they allow specializing fault tolerance for a given application but they lack transparency for the developer.

There are more fully automatic fault tolerance middlewares. FT/MPI [Batchu et al., 2001] and P2P-MPI [Rattanapoka, 2008] provide fault tolerance based on redundancy. Processes are replicated and, thus, failure of a node with a replicated process will not affect the computation. These methods prevent service interruption but they target on platforms with a large number of nodes because they consume a lot of resources

CoCheck [Stellner, 1996] is one of the first solutions to add fault tolerance to MPI in 1996. It uses a blocking coordinated checkpoint/restart protocol to create a consistent global state of the application and Condor to checkpoint each process locally. Coordinated checkpoint/restart is the widely used fault tolerance protocol. It has been implemented and optimized in different variants for MPI in Starfish MPI [Agbaria and Friedman, 2003], LAM/MPI [Sankaran et al., 2005], MPICH-Cl [Bouteiller et al., 2003b], MPICH-VCl [Lemarinier et al., 2004], MPICH-Pcl [Coti et al., 2006], Open MPI [Hursey et al., 2007] and in other programming models in Charm++ [Zheng et al., 2004] and in Kaapi with CCK [Besseron and Gautier, 2008]

An uncoordinated checkpoint/restart protocol is also implemented in Starfish MPI [Agbaria and Friedman, 2003]. It does not suffer from the domino effect since it uses atomic group communications that ensure reliable ordered message delivery. But uncoordinated checkpoint/restart protocols are mostly used with message logging protocols.

Communication induced checkpointing protocols have been experimented with Egida [Alvisi et al., 1999]; and a specialized version for work stealing is implemented in Kaapi with the TIC (Thief Induced Checkpointing) [Jafar et al., 2009].

Message logging protocols have also been widely studied. Egida [Rao et al., 2000] offers its own language to express the different message logging protocols: pessimistic, optimistic and causal. Some implementation of pessimistic message logging protocols is proposed in MPI-FT [Louca et al., 2000], MPICH-V1 [Bosilca et al., 2002], MPICH-V2 [Bouteiller et al., 2003a] and Charm++ [Chakravorty and Kalé, 2004]. Causal message logging protocols have been experimented through Manetho [Elnozahy, 1993] and MPICH-VCausal [Lemarinier et al., 2004].

The Charm++ middleware offers automatic load balancing at run-time. Thanks to this feature, Charm++ requires no spare processors as the failed processes states will be restored on non-failed processes and load balanced. Adaptive MPI (AMPI) is an MPI implementation on top of Charm++, so it automatically benefits from load balancing and the fault tolerance properties of Charm++.

A Kaapi application is described as a data flow graph which is distributed among the processors using a work stealing scheduler. Fault tolerance protocols implemented in Kaapi, TIC (Thief Induced Checkpointing) [Jafar et al., 2009] and CCK (Coordinated Checkpointing in Kaapi) [Besseron and Gautier, 2008] are specialized versions of classical protocols for the data flow graph model. This allows many optimizations like recovering without a spare processor or determining finely lost computations in case of failure.

Finally, Satin [Wrzesinska et al., 2006] is a middleware that offers fault tolerance services through an original approach that is not based on the classical methods like checkpoint or message logging, but it is specialized for the work stealing scheduling [Wrzesinska et al., 2006]. In case of failure, the results from orphan jobs (i.e., jobs stolen from a failed processor) are stored in a global table. The jobs stolen by a failed processor are rescheduled and they can use results from the global table instead of recomputing them.

This large variety of implemented protocols is rarely compared to each other. Exceptions are inside some middlewares that implement different variants like Egida [Rao et al., 2000] and MPICH-V [Bouteiller et al., 2006], [Lemarinier et al., 2004], [Bouteiller et al., 2003b], [Coti et al., 2006].

Open MPI is an implementation that supports the entire MPI-2 standard. Its fault tolerance architecture was designed to be flexible and modular in order to encourage the experimentation of alternative techniques. Its purpose is to provide an unified approach for MPI. It is split in five components [Hursey et al., 2007]. *Snapshot Coordinator* is responsible for launching, monitoring and the aggregation of checkpoint requests. *File Management* manages checkpoint related files and directories. *Distributed Checkpoint/Restart Coordination Protocol* is the coordination protocol that guarantees the distributed state's consistency. Currently, only a coordinated checkpoint/restart protocol similar to the one used in LAM/MPI is implemented. *Local Checkpoint/Restart System* is responsible for checkpointing or restoring the local process's state. At this time, it supports an application level snapshot using callbacks offered by the API and a memory space based snapshot using the Berkeley Lab's Linux Checkpoint/Restart library. *MPI Library Notification Mechanisms* notifies and coordinates the subsystems of the MPI implementation for checkpoint/restart events.

6.6.3   Implementation comparison

This section details some recent fault tolerant middlewares. They were chosen because they are widely used or because they propose an original approach of the fault tolerance problem. They should provide a representative overview of the current solutions for automatic fault tolerance. They are all compared in the Table 6.1 using the following criteria. *Single process snapshot* states the methods used to take a process snapshot as described in Section 6.6.1; this influences directly the portability and the size of the checkpointed state. *Protocol* corresponds to the fault tolerance protocol used; they are detailed in Section 6.4. *Storage component* refers to the physical component that is used for storing checkpointed states or logged messages. *Reliable components* states which components are supposed to be reliable in that implementation. *Spare processor* indicates if a spare processor is required for restarting the application. Restarting without spare processors requires a load balancing system that prevents performance falls. *Recovery* defines how many processes rollback to a previous state: Global for all, Local for only the

Middleware	Single process snapshot	Protocol	Storage component	Reliable components	Spare processor	Recovery
CoCheck (1996)	Memory space (Condor)	Blocking coordinated checkpointing	Checkpoint server	Stable memory	Required	Global
MPICH-Cl (2003)	Memory space (Condor)	Non-blocking coordinated checkpointing	Checkpoint server	Checkpoint Servers + 1 Dispatcher + 1 Checkpoint Scheduler	Required	Global
MPICH-Vcl (2003)	Memory space (Condor, libckpt or BLCR)	Non-blocking coordinated checkpointing	Local node + Checkpoint server	Checkpoint Servers + 1 Dispatcher + 1 Checkpoint Scheduler	Required	Global
FTC-Charm++ (2004)	Middleware based	Blocking coordinated checkpointing	Local node + Buddy processor	-	Not required	Global
MPICH-Pcl (2006)	Memory space (Condor, libckpt or BLCR)	Blocking coordinated checkpointing	Local node + Checkpoint server	Checkpoint Servers + 1 mpiexec process	Required	Global
Open MPI (2007)	Memory space (BLCR) or application level	Blocking coordinated checkpointing	Checkpoint server	Checkpoint servers	Required	Global
Kaapi-CCK (2008)	Middleware based	Blocking coordinated checkpointing	Checkpoint server	Checkpoint servers + 1 Master node	Not required	Partial
Kaapi-TIC (2005)	Middleware based	Communication induced checkpointing	Checkpoint server	Checkpoint servers + 1 Master node	Not required	Local
MPICH-V1 (2002)	Memory space (Condor)	Pessimistic message logging	Checkpoint server + Channel Memory	Checkpoint Servers + Channel Memories + 1 Dispatcher	Required	Local
MPICH-V2 (2003)	Memory space (Condor)	Sender-based pessimistic message logging	Checkpoint server + Event Logger	Checkpoint Servers + Event loggers + 1 Dispatcher + 1 Checkpoint Scheduler	Required	Local
MPICH-VCausal (2004)	Memory space (Condor)	Sender-based causal message logging	Checkpoint server + Event Logger	Checkpoint Servers + Event loggers + 1 Dispatcher + 1 Checkpoint Scheduler	Required	Local
FTL-Charm++ (2004)	Middleware based	Sender-based pessimistic message logging	Local node + Buddy processor	-	Required	Local
Satin (2006)	-	Ad hoc protocol using a global table	-	-	Not required	-

Table 6.1: Summary of main fault tolerance protocol implementations.

failed processes, and Partial for intermediate solutions.

6.7   Concluding remarks

The goal of this chapter was to present the most important models and protocols for designing efficient mechanisms that improve fault-tolerance in computational grids.

We started by describing an approach that looked at the problem as a multi-objective problem. The idea is to optimize the resource utilization by way of a classical scheduling problem and add a reliability function as an extra objective. Optimizing both objectives together corresponds to determine a good trade-off between performance and reliability. Then, the system administrator acts as a decision maker and chooses one trade-off among a set of "good" trade-offs. We discussed a couple of variants for solving these bi-objective problems depending on whether task duplication is allowed, and whether faults are permanent. Duplication may improve drastically the reliability; however, it leads to hard intractable problems.

Another approach is to add periodically a checkpoint and store the intermediate states on an external and safe memory, in order to be able to recover when a failure occurs and restart the computations from the last checkpoint. We described in details the main existing checkpoint protocols and the main available implementations. We also provided a survey of the large number of stochastic checkpoint models. The goal of all these models is to optimize a trade-off between the amount of work lost when a failure happens and the overhead due to checkpointing. Most models express these trade-offs as a cost function which reflects the mean completion time of the applications. This way, it is relatively easy to address a large variety of problems.

The purpose of this chapter was not to present hard technical problems in detail; however, we hope that it allows to provide a deep vision of problems and techniques for dealing with faults in computational grids, especially those relying on checkpoints. Several challenging open research directions can be stressed: first, it is interesting to refine more the theoretical stochastic models for taking into account more features. Of course, the more sophisticated the model, the harder the resolution, in particular, only numerical resolutions could be applied. The second direction would be to use such models for solving variants of the problem, like for computing the optimal interval length between two consecutive checkpoints (it may depend on the structure of the application), or computing the minimum number of checkpoint servers for saving the intermediate states. Third, a more difficult point would be to improve the time for checkpointing, for instance in considering decentralized strategies. This becomes crucial with the growth of the number of processors in computational grids. Another way is to investigate the smart encoding of the data to checkpoint. Finally, in a more algorithmic setting, it is worthwhile to study how to design algorithms that are intrinsically self-tolerant.

6.8   References

[Adiga, 2002] Adiga, N. (2002). An overview of the BlueGene/L supercomputer. In *Proceedings of Supercomputing*, pages 1–22.

[Agbaria and Friedman, 2003] Agbaria, A. and Friedman, R. (2003). Starfish: fault-tolerant dynamic MPI programs on clusters of workstations. *Cluster Computing*, 6(3):227–236.

[Alvisi and Marzullo, 1998] Alvisi, L. and Marzullo, K. (1998). Message logging: pessimistic, optimistic, causal, and optimal. *IEEE Transactions on Software Engineering*, 24(2):149–159.

[Alvisi et al., 1999] Alvisi, L., Rao, S., Husain, S., Mel, A., and Elnozahy, E. (1999). An analysis of communication-induced checkpointing. *International Symposium On Fault-Tolerant Computing*, pages 242–249.

[Assayad et al., 2004] Assayad, I., Girault, A., and Kalla, H. (2004). A bi-criteria scheduling heuristic for distributed embedded systems under reliability and real-time constraints. In *Proceedings of DSN*, pages 347–356.

[Aulwes et al., 2004] Aulwes, R., Daniel, D., Desai, N., Graham, R., Risinger, L., Taylor, M., Woodall, T., and Sukalski, M. (2004). Architecture of LA-MPI: a network-fault-tolerant MPI. In *Proceedings of Parallel and Distributed Processing*, page 15b. IEEE Computer Society.

[Avizienis et al., 2004] Avizienis, A., Laprie, J., Randell, B., and Landwehr, C. (2004). Basic concepts and taxonomy of dependable and secure computing. *IEEE Transactions on Dependable and Secure Computing*, 1(1):11–33.

[Baldoni, 1997] Baldoni, R. (1997). A communication-induced checkpointing protocol that ensures rollback-dependency trackability. In *Proceedings of FTCS*, page 68. IEEE Computer Society.

[Barlow and Proschan, 1996] Barlow, R. E. and Proschan, F. (1996). *Mathematical theory of reliability*. SIAM, classics edition.

[Batchu et al., 2001] Batchu, R., Skjellum, A., Cui, Z., Beddhu, M., Neelamegam, J., Dandass, Y., and Apte, M. (2001). MPI/FT: architecture and taxonomies for fault-tolerant, message-passing middleware for performance-portable parallel computing. In *Proceedings of Cluster Computing and the Grid*, page 26. IEEE Computer Society.

[Benoit et al., 2008] Benoit, A., Hakem, M., and Robert, Y. (2008). Fault tolerance scheduling of precedence task graphs on heterogeneous platforms. In *Proceedings of IEEE International Parallel and Distributed Processing Symposium*, pages 1–8.

[Besseron and Gautier, 2008] Besseron, X. and Gautier, T. (2008). Optimised recovery with a coordinated checkpoint/rollback protocol for domain decomposition applications. In *Proceedings of MCO*, pages 497–506.

[Bosilca et al., 2002] Bosilca, G., Bouteiller, A., Cappello, F., Djilali, S., Fedak, G., Germain, C., Hérault, T., Lemarinier, P., Lodygensky, O., Magniette, F., Néri, V., and Selikhov, A. (2002). MPICH-V: toward a scalable fault tolerant MPI for volatile nodes. In *Proceedings of SuperComputing*, page 29. IEEE Computer Society.

[Bouteiller et al., 2003a] Bouteiller, A., Cappello, F., Hérault, T., Krawezik, G., Lemarinier, P., and F.Magniette (2003a). MPICH-V2: a fault tolerant MPI for volatile nodes based on pessimistic sender based message logging. In *Proceedings of SuperComputing*, page 25. IEEE Computer Society.

[Bouteiller et al., 2006] Bouteiller, A., Hérault, T., Krawezik, G., Lemarinier, P., and Cappello, F. (2006). MPICH-V project: a multiprotocol automatic fault tolerant MPI. *The International Journal of High Performance Computing Applications*, 20:319–333.

[Bouteiller et al., 2003b] Bouteiller, A., Lemarinier, P., Krawezik, G., and Cappello, F. (2003b). Coordinated checkpoint versus message log for fault tolerant MPI. In *Proceedings of Cluster Computing*, page 242. IEEE Computer Society.

[Chakravorty and Kalé, 2004] Chakravorty, S. and Kalé, L. (2004). A fault tolerant protocol for massively parallel systems. *Proceeedings of Parallel and Distributed Processing*, 12:212a.

[Chandy and Lamport, 1985] Chandy, K. and Lamport, L. (1985). Distributed snapshots: determining global states of distributed systems. *ACM Transactions on Computer Systems*, 3(1):63–75.

[Chandy and Ramamoorthy, 1972] Chandy, K. and Ramamoorthy, C. (1972). Rollback and recovery strategies for computer programs. *IEEE Transactions on Computers*, 21(6):546–556.

[Chimento and Trivedi, 1993] Chimento, P. and Trivedi, K. (1993). The completion time of programs on processors subject to failure and repair. *IEEE Transactions on Computers*, 42(10):1184–1194.

[Coti et al., 2006] Coti, C., Hérault, T., Lemarinier, P., Pilard, L., Rezmerita, A., Rodriguez, E., and Cappello, F. (2006). Blocking versus non-blocking coordinated checkpointing for large-scale fault tolerant MPI. In *Proceedings of SuperComputing*, page 18. IEEE Computer Society.

[Crutchfield et al., 2008] Crutchfield, C. Y., Dzunic, Z., Fineman, J. T., Karger, D. R., and Scott, J. H. (2008). Improved approximations for multiprocessor scheduling under uncertainty. In *Proceedings of SPAA*, pages 246–255.

[Daly, 2006] Daly, J. (2006). A higher order estimate of the optimum check-point interval for restart dumps. *Future Generation Computer Systems*, 22(3):303–312.

[Dogan and Özguner, 2002] Dogan, A. and Özguner, F. (2002). Matching and scheduling algorithms for minimizing execution time and failure probability of applications in heterogeneous computing. *IEEE Transactions on Parallel and Distributed Systems*, 13(3):308–324.

[Dongarra et al., 2007] Dongarra, J., Jeannot, E., Saule, E., and Shi, Z. (2007). Bi-objective scheduling algorithms for optimizing makespan and reliability on heterogeneous systems. In *Proceedings of SPAA*, pages 280–288.

[Duell et al., 2002] Duell, J., Hargrove, P., and Roman, E. (2002). The design and implementation of Berkeley lab's Linux checkpoint/restart. Technical report, Lawrence Berkeley National Laboratory. Available online at: http://repositories.cdlib.org/lbnl/LBNL-54941 (accessed May 1, 2009).

[Elnozahy, 1993] Elnozahy, E. (1993). *Manetho: fault tolerance in distributed systems using rollback-recovery and process replication*. PhD thesis, Rice University, Houston, TX, USA.

[Elnozahy et al., 2002] Elnozahy, E., Alvisi, L., Wang, Y.-M., and Johnson, D. (2002). A survey of rollback-recovery protocols in message-passing systems. *ACM Comput. Surv.*, 34(3):375–408.

[Elnozahy and Plank, 2004] Elnozahy, E. and Plank, J. (2004). Checkpointing for peta-scale systems: a look into the future of practical rollback-recovery. *IEEE Transactions on Dependable and Secure Computing*, 1(2):97–108.

[Fagg et al., 2001] Fagg, G., Bukovsky, A., and Dongarra, J. (2001). HARNESS and fault tolerant MPI. *Parallel Computing*, 27(11):1479–1495.

[Girault et al., 2009] Girault, A., Saule, E., and Trystram, D. (2009). Reliability versus performance for critical applications. *Journal of Parallel and Distributed Computing*, 69(3):326–336.

[Hoogeveen, 2004] Hoogeveen, H. (2004). Multicriteria scheduling. *European Journal of Operational Research*, 167(3):592–623.

[Huang and Abraham, 1984] Huang, K. and Abraham, J. (1984). Algorithm-based fault tolerance for matrix operations. *IEEE Transactions on Computers*, 33(6):518–528.

[Hursey et al., 2007] Hursey, J., Squyres, J., Mattox, T., and Lumsdaine, A. (2007). The design and implementation of checkpoint/restart process fault tolerance for Open MPI. In *Proceedings of IEEE International Parallel and Distributed Processing Symposium*. IEEE Computer Society.

[Jafar et al., 2009] Jafar, S., Krings, A., and Gautier, T. (2009). Flexible rollback recovery in dynamic heterogeneous grid computing. *IEEE Transactions on Dependable and Secure Computing*, 6(1):32–44.

[Jeannot et al., 2008] Jeannot, E., Saule, E., and Trystram, D. (2008). Bi-objective approximation scheme for makespan and reliability optimization on uniform parallel machines. In *Proceedings of the Euro-Par Conference*, pages 877–886.

[Leighton and Ma, 1999] Leighton, F. and Ma, Y. (1999). Tight bounds on the size of fault-tolerant merging and sorting networks with destructive faults. *SIAM Journal of Computing*, 29(1):258–273.

[Lemarinier et al., 2004] Lemarinier, P., Bouteiller, A., Hérault, T., Krawezik, G., and Cappello, F. (2004). Improved message logging versus improved coordinated checkpointing for fault tolerant MPI. In *Proceedings of CLUSTER*, pages 115–124. IEEE Computer Society.

[Leung and Choo, 1984] Leung, C. and Choo, Q. (1984). On the execution of large batch programs in unreliable computing systems. *IEEE Transactions on Software Engineering*, 10(4):444–450.

[Leung, 2004] Leung, J. (2004). *Handbook of scheduling: algorithms, models, and performance analysis*. Chapman & Hall/CRC Press, Boca Raton, FL, USA.

[Liang et al., 2006] Liang, Y., Zhang, Y., Sivasubramaniam, A., Jette, M., and Sahoo, R. (2006). BlueGene/L failure analysis and prediction models. In *Proceedings of DSN*, pages 425–434. IEEE Computer Society.

[Litzkow et al., 1997] Litzkow, M., Tannenbaum, T., Basney, J., and Livny, M. (1997). Checkpoint and migration of Unix processes in the Condor distributed processing system. Technical report, University of Wisconsin-Madison, Madison, WI, USA.

[Lloyd and Lipow, 1962] Lloyd, D. and Lipow, M. (1962). *Reliability: management, methods, and mathematics*, chapter 9. PrenticeHall, Englewood Cliffs, NJ, USA.

[Louca et al., 2000] Louca, S., Neophytou, N., Lachanas, A., and Evripidou, P. (2000). MPI-FT: portable fault tolerance scheme for MPI. *Parallel Processing Letters*, 10(4):371–382.

[Mahjoub et al., 2009] Mahjoub, A., Pecero, J., and Trystram, D. (2009). Scheduling with uncertainties on new computing platforms. *Computational Optimization and Applications*.

[Oliner et al., 2004] Oliner, A., Sahoo, R., Moreira, J., Gupta, M., and Siva-subramaniam, A. (2004). Fault-aware job scheduling for BlueGene/L systems. In *Proceedings of IEEE International Parallel and Distributed Processing Symposium*, page 64. IEEE Computer Society.

[Plank et al., 1995] Plank, J., Beck, M., Kingsley, G., and Li, K. (1995). Libckpt: transparent checkpointing under Unix. In *Usenix Winter Technical Conference*, pages 213–223. Available online at: http://www.cs.utk.edu/~plank/plank/papers/papers.html (accessed May 1, 2009).

[Randell, 1975] Randell, B. (1975). System structure for software fault tolerance. In *Proceedings of Reliable Software*, pages 437–449.

[Rao et al., 2000] Rao, S., Alvisi, L., and Vin, H. (2000). The cost of recovery in message logging protocols. *IEEE Transactions on Knowledge and Data Engineering*, 12(2):160–173.

[Rattanapoka, 2008] Rattanapoka, C. (2008). *P2P-MPI: a fault-tolerant message passing interface implementation for grids*. PhD thesis, Université Louis Pasteur, Strasbourg.

[Sankaran et al., 2005] Sankaran, S., Squyres, J., Barrett, B., Lumsdaine, A., Duell, J., Hargrove, P., and Roman, E. (2005). The LAM/MPI checkpoint/restart framework: system-initiated checkpointing. *International Journal of High Performance Computing Applications*, 19(4):479–493.

[Saule and Trystram, 2009] Saule, E. and Trystram, D. (2009). Analyzing scheduling with transient failures. *IPL*, 109(11):539–542.

[Schroeder and Gibson, 2006] Schroeder, B. and Gibson, G. (2006). A large-scale study of failures in high-performance computing systems. In *Proceedings of DSN*, pages 249–258.

[Siewiorek and Swarz, 1998] Siewiorek, D. and Swarz, R. (1998). *Reliable computer systems, design and evaluation*. A.K. Peters, 3rd edition.

[Stellner, 1996] Stellner, G. (1996). CoCheck: checkpointing and process migration for MPI. In *Proceedings of the International Symposium on Parallel Processing*, page 526. IEEE Computer Society.

[Strom and Yemini, 1985] Strom, R. and Yemini, S. (1985). Optimistic recovery in distributed systems. *ACM Transaction on Computer Systems*, 3(3):204–226.

[Tamir and Séquin, 1984] Tamir, Y. and Séquin, C. (1984). Error recovery in multicomputers using global checkpoints. In *Proceedings of the 1984 International Conference On Parallel Processing*, pages 32–41. Available online at: http://www.cs.ucla.edu/~tamir/papers/icpp84.pdf (accessed May 1, 2009).

[Toueg and Babaoglu, 1983] Toueg, S. and Babaoglu, O. (1983). On the optimum checkpoint selection problem. *SIAM Journal on Computing*, 13:630–649.

[Wrzesinska et al., 2006] Wrzesinska, G., Van, R. N., Maassen, J., Kielmann, T., and Bal, H. (2006). Fault-tolerant scheduling of fine-grained tasks in grid environments. *International Journal of High Performance Computing Applications*, 20(1):103–114.

[Young, 1974] Young, J. (1974). A first order approximation to the optimum checkpoint interval. *Communications of ACM*, 17(9):530–531.

[Zheng et al., 2004] Zheng, G., Shi, L., and Kalé, L. (2004). FTC-Charm++: an in-memory checkpoint-based fault tolerant runtime for Charm++ and MPI. In *Proceedings of Cluster Computing*, pages 93–103. IEEE Computer Society.

[Ziv and Bruck, 1997] Ziv, A. and Bruck, J. (1997). An on-line algorithm for checkpoint placement. *IEEE Transactions on Computers*, 46(9):976–985.

Chapter 7

Fault tolerance for distributed scheduling in grids

Lei Yu

Applied Mathematics and Systems Laboratory, Ecole Centrale Paris, Grande Voie des Vignes, 92295 Châtenay-Malabry, France

Frédéric Magoulès

Applied Mathematics and Systems Laboratory, Ecole Centrale Paris, Grande Voie des Vignes, 92295 Châtenay-Malabry, France

7.1 Introduction

Along with the deployment of more and more heterogeneous clusters, grid computing has become an increasingly popular solution for leveraging existing IT infrastructure to optimize computing resources and manage data and computing workloads. Lots of grid projects have been launched to build a national problem-solving system on the grid, such as GrADS [Berman et al., 2001] and DIET [Caron and Desprez, 2006]. These projects aim to connect the nation's computers, databases and instruments in a seamless grid, supporting emerging computation-rich application concepts such as remote computing, distributed supercomputing, tele-immersion, smart instruments and data mining. In these large scale systems, the scheduling and fault tolerance are obviously key technical obstacles to be overcome. According to the presentation of Hamscher and his colleagues [Hamscher et al., 2000], the meta-scheduling architecture can be included into three principal schemas: centralized scheduling, hierarchical scheduling and distributed scheduling. The main advantage for hierarchical and distributed scheduling is the fact that different policies can be used for local and global job scheduling, the communication bottleneck of centralized scheduling is prevented and the system is more scalable. But in the hierarchical and distributed structure, each resource has its own administrative domain. These resources are geographically distributed and are gathered using a WAN or even Internet. Those characteristics lead these scheduling structures to be more error prone than other computing environments. A fault tolerant mechanism should be proposed to detect automatically the failure of components and to ensure that the failure will not affect the whole grid system.

At present, computational grids and grid services have become an important asset in large-scale scientific and engineering research. Many scientific communities are feeling a growing need to integrate their legacy applications into grid environments. By wrapping command-line applications into grid services and scheduling these grid services for end-users, a framework has been implemented in [Yu and Magoulès, 2007] to enable the dynamic deployment, the discovering and the submission of scientific applications in a grid environment. In order to integrate and schedule large-scale distributed applications, a scheduling structure which is not centralized is more appreciated. Thus a fault tolerant scheduling algorithm must be realized in the framework to increase the system stability as we have mentioned before.

Thus the target system can be considered as a tree structure of schedulers which are organized in a hierarchical manner. The problem which must be resolved in this chapter is concluded as follows:

- A distributed scheduling algorithm to achieve the effective and fault tolerant resource discovery even if one of the schedulers in the tree structure fails or the transferred messages are lost.

- A mechanism to detect the failure of schedulers in the hierarchical structure and reconstruct automatically the tree structure of schedulers to enable the sub-tree and components of the failed scheduler.

The rest of this chapter is as follows. Fault tolerance theory and technology are presented in Section 7.2. Then the distributed scheduling model architecture is described and the fault tolerance issues for each component are discussed in Section 7.3. The detail of fault detection and repairing is presented in Section 7.4. In Section 7.5 the fault tolerant distributed scheduling algorithm (DDFT) is described. SimGrid and our simulation application are presented in Section 7.6, and several simulations are conducted to evaluate the performance of the proposed model in Section 7.7. The related work is discussed in Section 7.8. Finally we conclude with a brief discussion of the future research.

7.2 Fault tolerance in distributed systems

Before the presentation of our fault tolerant model, theories of fault tolerance are introduced to facilitate the understanding of terms used later in the chapter.

Fault tolerance is an ability of a system to respond gracefully to an unexpected hardware or software failure. A distributed system has the partial failure property. Because of the dispersion of processing resources in the distributed system, no matter what kind of failure occurs, it usually affects only a part of the entire system. Thus the fault tolerance in a distributed system is considered as the possibility that the tasks of failing processes can be taken over by the remaining components, leading to a graceful degradation rather than an overall malfunctioning.

Two principal technologies can be used to achieve the fault tolerance of systems: stable memory and replication processes. In stable memory, recovery points are regularly saved by processes. When a process fails, it is recovered from the last recovery point. However for replication processes, a replication of process or a backup server is implemented to increase the reliability of a distributed system.

In a distributed system, fault tolerance becomes considerably more difficult for distributed applications, made up of several processes that communicate by passing messages between themselves. Because of the partial failure property of a distributed system, it would be a great challenge that the global state of a distributed application is kept consistent when one process fails. An application is in a globally consistent state if whenever the receipt operation of a message has been recorded in the state of some process, then the send operation of that message must have been recorded also [Dialani et al., 2002]. Precisely, for the stable memory technology, the recovery point of each process should be consistent. Three strategies can be used to save recovery points in order to keep the consistency of a global state.

- The process saves recovery points in an asynchronous way, without the dependence of the other processes. In the case of failure, we need a set of recovery points which represent a coherent global state of a system to restart the computing.

- The recording of the recovery points is preset in order to represent correspondingly a coherent global state. There should be many messages exchanged between processes.

- Dynamic coordination between the actions of recording of recovery points.

In the case of replication processes, the problem is to keep the consistency among replications of process or back servers. Formally three strategies ensure that the state of each backup is identical: passive replication, active replication and semi-active replication. Passive replication distinguishes two behaviors from replications: the primary copy and backups. The primary copy is the only one to carry out all the treatments. The backups supervise passively the primary copy. In the case of failure of the primary copy, a backup becomes the new primary copy. Active replication treats equally each backup of process. All the backups receive the same sequence, and all the backups process the request independently. After the treatment of the request, the backups send the result to the client autonomously. Then the client can select a response as the result. Semi-active replication is located between the active replication and the passive replication. All the backups receive the same sequence, and all the backups process the request independently. But after the treatment of the request, the primary copy is the only one to be sent as the result.

7.3 Distributed scheduling model

The distributed scheduling model proposed has a tree structure. The root of the tree is master meta-scheduler (MMS) which provides a portal to interact with clients. The leaves of the tree are computing resources (CR) which encapsulate scientific applications and execute users jobs. The CRs are grouped into sites. For each site, a site meta-scheduler (SMS) is deployed to make the local resource mapping for user jobs. Local meta-scheduler (LMS) is the node between the root and SMSs. It is needed for a large-scale scheduling system and can be optional according to the system structure and scale. Each tree in the structure represents a virtual organization (VO) in the grid. According to the definition in the paper [Foster et al., 2001], VO is a set of individuals and/or institutions defined by sharing rules in order to share their information and direct access to computers, software, data and other resources. These sharing rules define clearly and carefully just what is shared, who is allowed to share, and the conditions under which sharing occurs. The model structure is shown in Figure 7.1 and the fault tolerance of each component in this structure is discussed as follows.

7.3.1 MMS fault tolerance

MMS is the key component in the model. It is the only entrance for clients to practice the grid and it provides a friendly graphic user interface (GUI) portal to facilitate application request, job submission and job monitoring. Therefore the execution of MMS must be ensured by fault tolerant technologies. According to the introduction of Section 7.2, the passive replication should be an appropriate solution. Each MMS has a backup server which

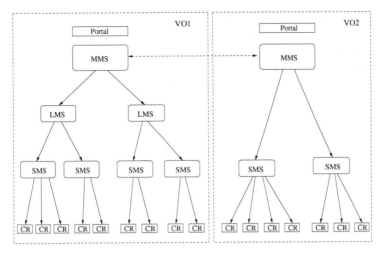

FIGURE 7.1: The distributed scheduling structure.

supervises passively the MMS. In the case of failure of MMS, the backup becomes the new MMS. Thus recovery points must be saved regularly by MMS and the execution state should be recovered by the new MMS when the MMS fails.

The problem that must be resolved here is the communication between the new MMS and LMSs/SMSs. When a MMS breaks down, how can LMSs/SMSs find out the new MMS? One solution should be the configuration file. Each child of MMS has a configuration file which can be used to set backup server information, e.g., IP address. When its child can not receive response of MMS anymore in a predefined delay, it can try to connect the backup server.

7.3.2 LMS/SMS fault tolerance

SMS manages the CRs in a site and makes scheduling decisions in a site level. Located in the middle of MMS and SMS, LMS takes concurrently the responsibility of transferring messages between MMS and SMSs. The quantity of LMS and SMS is larger than MMS, and the deployment of LMS and SMS should be more flexible and portable in order to make the VO system more scalable. Thus the solution of replication is not suitable for the case of LMS/SMS. A series of algorithms are proposed to detect automatically the failure of LMSs or SMSs, to achieve the reconstruction of the connection between child nodes of crashed LMS/SMS and MMS (LMS) in the case of malfunction of a LMS/SMS and to realize the restart and recovery of the crashed LMS/SMS.

The principle of these algorithms is message exchanges in the tree structure. Through these messages, each component of our model can acquire knowledge of system topologies and can detect the malfunction of the component from which it receives messages. If a component doesn't receive messages from its parent node in a predefined delay, it considers that its parent node may be malfunctioned. Then the component sends messages to its parent to verify the failure. If it confirms that its parent is malfunctioned, the component tries to reconstruct the connection with other components which situate at the level of its parent or the higher level, e.g., its grandparent, because it has the knowledge of whole system topologies.

7.3.3   CR fault tolerance

Normally, a CR is a cluster of computers which has a head node to manage and schedule jobs onto each node. Thus the failure of CR can be concluded into two types of failure: the head node failure and the slave nodes failure. The head node failure is considered as a single point of failure (SPoF) which causes the cluster to go down completely. Being inspired by the solution [Limaye et al., 2005b], we propose a solution to resolve the problem of head node failure. A service (monitoring service) which is similar with the HA-OSCAR monitoring service [Limaye et al., 2005b] is deployed in site scheduler (SMS). This service attempts to restart the critical services in the head node if these services have failed. In the worst case where the monitoring service can not restart the failed service after several attempts, this CR is unavailable and all the jobs running in this CR should be rescheduled by site scheduler. The site scheduler first tries the other CRs in the site to restart these jobs. For the jobs which can not be restarted in the site, they will be sent to MMS by the site scheduler. MMS maintains a rescheduling job queue, and regularly maps jobs in this queue to available CRs.

The problem of the slave node failure strongly depends on the local resource manager system (e.g., Condor [Thain et al., 2005], PBS [Bayucan et al., 1999]) of each site. An important scheme to achieve fault tolerance, for running jobs in cluster systems, is checkpoint/restart technique. Checkpointing is a procedure of storing process state to a file, which is later used to reconstruct the process. Several researchers [Limaye et al., 2005a], [Limaye et al., 2005c], [Gracjan et al., 2006] aim to add fault tolerant mechanisms in a cluster and restart a job of failed nodes from the last checkpoint in another slave node using its checkpoint file. Therefore with such mechanisms, jobs will be automatically completed in another node even if its original execution node fails.

7.4 Fault detection and repairing in the tree structure

7.4.1 Notations

In order to clearly explain the mechanism of fault tolerance, several messages which are exchanged between components are defined. Local status publication (LSP) is made by a node in the tree (LMS, SMS or CR) and contains topology information of the sub-tree of this node. This message is sent by a node to its parent node when this node wants to participate in the VO or when its local topology is changed. Finally, the LSP message is collected by the MMS in a VO with the aggregation of the message between LMSs/MSMs. Having the whole topology knowledge of VO, the MMS then makes the message complete topology publication (CTP) which contains the topology information of the VO and sends regularly CTP to its child nodes in the manner of a *heartbeat*. Another very important message is "is-alive" which is used by a component to verify the malfunction of its parent node. The transfer direction and path of messages are shown in the Figure 7.2.

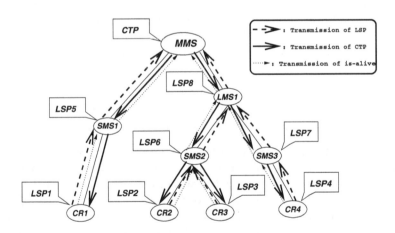

FIGURE 7.2: The message transfer model.

7.4.2 Algorithms description

7.4.2.1 Message treatment algorithms

The fault tolerance in the distributed scheduling model is achieved by exchanging messages between components. Therefore, the treatment of messages in each node is the first problem that must be resolved. For the MMS, there are four types of messages which need to be treated: a node wants to be its child node, one of its child nodes updates the local topology and sends it a message LSP, the creation and delivery of message CTP and the response to message "is-alive." As shown in Algorithm 7.4.1, if MMS receives a message LSP, it first decides where the message comes from. If this message comes from its child, it means that its child updates the local topology. MMS then updates its CTP and sends the new CTP to all its child nodes. Otherwise, it is another node which wants to take part in the VO. In this case, MMS responds "OK" to the sender and adds the sender in the set of its child nodes. If a component wants to verify its status with message "is-alive," MMS must respond "OK" to confirm its functionality. Finally, MMS must deliver the message CTP regularly to all its child nodes. The advantages to deliver message CTP frequently can be explained as follows:

- Notification of whole system topology. Thus all the components in the system have the knowledge of system topology and the knowledge can be used to reconstruct the connection when a failure occurs.

- Consistency of CTP in each component. CTP is sent by local network or Internet which have possibilities to lose transferred messages. In the same time, the topology of sub-tree may also be changed. The frequent delivery of message CTP can assure the consistency of CTP even if some of messages CTP transferred over the network are lost or the local topology is changed.

- Fault tolerance. The interval of delivering message CTP is predefined. Thus if a component can not receive the message CTP from its parent node, it may consider that its parent node is malfunctioned.

For the case of LMS/SMS, Algorithm 7.4.2, the treatment of messages LSP and message "is-alive" is similar as in the case of MMS. LMS/SMS distinguishes the sender of message LSP: from its child or not. Then LMS/SMS must send the new LSP to its parent every time it updates its local LSP. CTP is received regularly from its parent node and it then updates its local version of CTP. Then the node delivers the received CTP to all its child nodes.

7.4.2.2 Algorithm for the failure detection and the connection reconstruction

The failure detection, Algorithm 7.4.3, can be achieved by the node's children. A delay is predefined and normally the node can receive the message CTP from its parent in the predefined delay. If a node does not receive a message CTP from its parent after this delay, it may think that its parent has malfunctioned. Thus a message "is-alive" is sent to its parent to verify the functionality of its parent node. If the node can not receive any responses from its parent, it makes the decision that its parent has failed. Therefore, it

---

Algorithm 7.4.1 Messages treatment in the MMS

var Sp : set of children

MMS : receive a message

if a message $< LSP, p >$ was received from p **then** {*treatment of message LSP*}

    if $p \notin Sp$ **then** {*LSP does not come from its child, another site which requires to take part*}

5:      send a message <OK, MMS> to p

      $Sp \Leftarrow Sp \cup \{p\}$

    end if

    modify CTP

    send a message $< CTP, g >$ to $\forall p \in Sp$

10: **end if**

if a message $< is\text{-}alive, p >$ was received from p **then** {*treatment of message is-alive*}

    send a message <OK, MMS> to p

end if

MA : send regularly CTP to $\forall p \in Sp$

---

requests another node to be its parent according to the knowledge of whole system topology and the predefined configuration, e.g., its grandparent. If the requested node responds "OK," the component replaces its parent with the requested node and the connection is reconstructed. Finally, the component receives regularly messages CTP from its new parent.

7.4.2.3 Recovery algorithm for the malfunctioned component

The malfunctioned node can be restarted soon. Thus the algorithm 7.4.4 is needed to recover the node from the failure. For the recovered node, two problems must be taken into account: how to find out and connect its former children and how to connect its former parent node. In order to resolve these problems, two messages are defined: recover sons publication (RSP) and request to leave (RL). Having saved recovery points regularly, the node has the knowledge of its former children and parent when it is recovered. Thus RSP is used by the recovered node to notify its former children that it has been recovered. Therefore the children which received the message RSP send the message RL to their current parent to request the departure. If their current parent responds "OK," they may send their local LSP to their former parent. After having recovered its former children, the recovered node then sends its new LSP to its former parent. If its former parent agrees, the new connection is reconstructed and the recovery is finished.

Algorithm 7.4.2 Messages treatment in the LMS/SMS

---

var Sp : set of children
var g : the process of node
node receives a message:
{*begin the treatment of messages*}
if a message < CTP, parent > was received **then** {*treatment of message CTP*}
5:   renew local CTP
     send a message $< CTP, g >$ to $\forall p \in Sp$
   end if
   if a message $< LSP, p >$ was received **then** {*treatment of message LSP*}
     if $p \notin Sp$ **then** {*LSP does not come from its child, another site which requires to take part*}
10:      send a message $< OK, g >$ to p
        $Sp \Leftarrow Sp \cup \{p\}$
     end if
     modify local LSP
     send a message $< LSP, g >$ to its parent
15: **end if**
   if a message $< is\text{-}alive, p >$ was received **then** {*treatment of message is-alive*}
     send a message $< OK, g >$ to p
   end if

Algorithm 7.4.3 Algorithm for detecting the failure and reconstructing the connection

---

var g : the process of node
if during $\Delta t1$ second, it does not receive message CTP **then** {*treatment of problem of failure*}
   send $< is\text{-}alive, g >$ to its parent several times
   if during $\Delta t2$ second, it does not receive message OK from its parent **then** {*the parent is broken down, tries to connect anther node p*}
     send $< LSP, g >$ to p
     if a message $< OK, p >$ was received from p **then**
       parent $= p$
     end if
   end if
end if

Algorithm 7.4.4 Algorithm of recovery

---

process p is restarted :
product a set of children Sp according to checkpoint
send message $< RSP, p >$ to $\forall p \in Sp$
waits $\Delta t1$
5: create LSP according to the children responses
repeat
    send a message $< LSP, p >$ to its former parent
    waits $\Delta t2$
until a message $< OK$, former parent $>$ was received
10: parent = former parent

for all components g who receive a message $< RSP, p >$:
send a message $< RL, g >$ to its parent
waits $\Delta t3$
if a message $< OK$, parent $>$ was received from its parent **then**
15:    send a message $< LSP, g >$ to p
    parent = p
end if

for all components f who receive a message $< RL, g >$:
send a message $< OK, f >$ to g
20: remove g from the set of children

7.4.3   Messages treatment analysis

The procedure of messages treatment in a node LMS or SMS, Algorithm 7.4.2, can be modeled by finite automata as illustrated in Figure 7.3. An automaton is defined by a quadruplet $A = <S, S_0, E, R>$ where: S is the set of states, S_0 is the initiate state, E is the set of events and R is the set of transitions. Therefore, for the automaton illustrated in the Figure 7.3, we can define: $S = \{S1, S2, \cdots, S12\}$, $S_0 = S1$, $E = \{E1, E2, \cdots, E12\}$ and $R \sqsubseteq S \times E \times S$.

In the automaton, all the states can be reached and there is not a state which has not a successor. The messages are transferred in the asynchronous manner; there is not a deadlock in the treatment and finally each treatment returns to the node's original state.

S1: Initiate state; S2: treatment of CTP ; S3: sending CTP; S4: treatment LSP comes from sons;
S5: treatment LSP from the others; S6: update LSP; S7: sending LSP; S8: find failure
S9: waiting father's response; S10: father's failure treatment.

E1: receive CTP; E2: send CTP; E3: sending completed; E4: receive LSP from sons;
E5: receive LSP from the others; E6: update local topology; E7: send LSP; E8: father's failure detect
E9: send "is–alive"; E10: response "OK"; E11: no response; E12: finish the treatment.

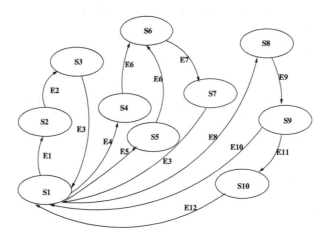

FIGURE 7.3: The automaton description of messages treatment.

7.5 Distributed scheduling algorithm

In Section 7.4, we have proposed a mechanism and a series of algorithms to achieve automatically the failed component detection and repairing. But the fault is only detected and repaired in a predefined delay; the scheduling algorithm must be robust enough to ensure the job submission to be effectuated correctly.

7.5.1 Distributed dynamic scheduling algorithm with fault tolerance (DDFT)

With the *resource service* deployed in CR, applications can be deployed and site scheduler (SMS) can submit jobs via its *resource factory* and *submission interface*.

Unlike the centralized scheme, the scheduling works are distributed into each SMS in this proposed model. At the beginning, the job request is submitted to the MMS by a client. Then, MMS delivers this request to all the site schedulers (SMS) through the tree structure. In each site, a resource is assigned to the job according to the local scheduling algorithm and strategy. Finally, the information of selected resource in each site is transferred to the MMS which makes a scheduling decision among the received information of resources according to its local scheduling strategy.

The DDFT scheduling algorithm which is realized in the model is described in Algorithm 7.5.1. When MMS receives a job request from a user, it sends this request to all its children. This request is transferred in the tree structure and finally it reaches each SMS. According to its local scheduling algorithm, a SMS makes a scheduling decision for the job, adds the job into its local job queue and returns the information of selected resource to its parent. When the parent (LMS or MMS) has collected all the responses of its children or it has waited a Δt duration, it selects the best resource for the job using the received responses. This treatment is repeated until MMS makes a scheduling decision for the job.

The procedure of submission is similar. The MMS sends a message "SUBMIT" associated with the information of selected resource to all its children. Then this message is transferred in the tree structure until it reaches each SMS. If the site has the selected resource, the job then is submitted to the resource by the SMS. Otherwise, the SMSs which do not have the selected resource delete the job from its local job queue.

When a job is completed, the SMS of the site where the job is executed sends a message "RESULT" to MMS. Then MMS sends the result to user and delivers a message "FINISHED" to all its children. This message will be transferred in the tree and the SMS which has the completed job removes the job from its local job queue when it receives the message "FINISHED."

Algorithm 7.5.1 DDFT scheduling algorithm
MMS receives a user job request (j)
MMS sends $SCHEDULING_j$ to all the sites
SMS receives $SCHEDULING_j$ and inserts j into the local job queue
SMS makes local scheduling decision (S_i) and sends to its parent
5:   MMS receives S_i and maps j to the best resource (r_j)
MMS sends submission request ($SUBMIT_j$) for j to its children
SMS receives $SUBMIT_j$
if $r_j \in$ the site (i) **then**
submit j to r_j
10:     send message $SUBMITTED_j$ to MMS
else
remove j from local job queue
end if
j is finished and SMS sends message $RESULT_j$ to MMS
15: MMS receives $RESULT_j$ and sends it to user
MMS sends message $FINISHED_j$ to all the sites
SMS receives $FINISHED_j$
if $j \in$ the local job queue **then**
remove j from job queue
20: **end if**

7.5.2   Algorithm fault tolerance issues

Two monitors are deployed to enhance the fault tolerance in our scheduling algorithm. In the MMS side, all the jobs which have been requested are maintained in a job queue and their status are monitored regularly. When MMS detects that a job has been scheduled or submitted a long time ago and MMS has not received the confirmed message for the job, MMS will reschedule or resubmit the job. Similarly, for a SMS, when it detects that a job in its local job queue has been completed a long time ago and it has not received the confirmed message from the MMS, it then re-sends the result to MMS. Therefore these two monitors can ensure that a job is successfully submitted and the result of job is finally sent to the client.

A use case is used to explain clearly the two monitors. In the Figure 7.2, we suppose that the SMS2 fails when the MMS submits the job j to the resource CR2. The message "SUBMIT" is sent to SMS1 and LMS1, then LMS1 sends this message to SMS2 and SMS3. SMS2 will not receive and submit j to CR2, because it has failed. Thus, j will not be executed and the message "SUBMITTED" will not be sent by SMS2. But with the monitor, MMS should find that j has been submitted and it has not been confirmed to be submitted by the resource in a predefined delay. Therefore the monitor re-sends the message "SUBMIT" to all the children of MMS for j. During this delay of re-sending a message, CR2 and CR3 should find the fault of its parent, because of the fault detecting mechanism, Section 7.4. We suppose that the new parent of CR2 and CR3 now is SMS3. Thus when MMS re-sends the message "SUBMIT" for j, SMS3 will receive and find that it has the job's requested resource. Then SMS3 submits j to the resource CR2.

We can suppose another possibility in this use case. CR2 completes a job j and sends the result to MMS when SMS2 fails. MMS can not receive the result because of the failure of SMS2. Then CR2 and CR3 detect the fault of their parent and find their new parent SMS3. The jobs which have been submitted to CR2 and CR3 are inserted into the local job queue of SMS3. Thus the monitor of SMS3 finds j has been completed a long time ago when it verifies job status. Therefore the monitor re-sends the result of j to its parent and the result reaches lastly MMS through LMS1. Then MMS sends the result to client and delivers a message "FINISHED" to all its children. When SMS3 receives the message "FINISHED" of j, it removes j from its local job queue.

7.6 SimGrid and simulation design

Our model describes a large-scale distributed scheduling system which is difficult to be tested and analyzed on real platforms. Consequently, one must resort to simulations, which enable reproducible results and also make it possible to explore wide ranges of platform and application scenarios.

The SimGrid project was initiated in 1999 to allow the study of scheduling algorithms for heterogeneous platforms. The SimGrid v3 which was released in 2005 adds many new features with respect to the previous versions. The simulation engine was completely rewritten, leading to better modularity, speed and scalability. Four user interfaces were supported to allow the use of the software in different contexts: SimDag, MSG, GRAS and SMPI [Casanova et al., 2008]. MSG is the user interface for researchers and is initially designed for studying scheduling algorithms. Thus the MSG interface is suitable for the simulation of our model.

Our simulation application is developed on SimGrid MSG interface and has a modular structure. The creation of platform and deployment of applications are separated from the programming of each module. In SimGrid, the description of platform and applications deployment is maintained separately in two files: platform.xml and deployment.xml. At the beginning of simulation, a platform constructor and an application deployer will use these two files to initiate the simulation environment. Thus the development of simulation system is decoupled from the real simulation environment, making the simulation system more independent and more suitable for multi-platform simulation.

Each module simulates the functionality of a component in our model. By exchanging messages, the modules can communicate with each other and can realize the job scheduling and fault detection. The structure of a simulation application is shown in Figure 7.4. According to the system configure files, SimGrid initiates the instances of components, e.g., LMS and SMS, and starts the simulation. The instances communicate with each other and create the tree of schedulers. The clients send requests of applications to MMS, and the requests will be transferred in the tree structure according to our DDFT algorithm. When a application is completed, the result is finally sent to the associated client.

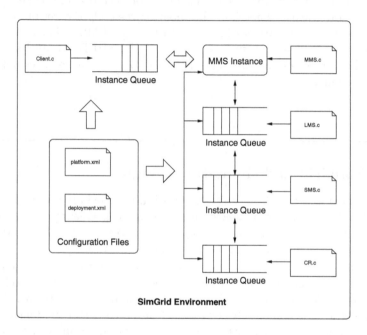

FIGURE 7.4: The simulation application structure.

7.7   Evaluation

In order to evaluate the performance of the proposed scheduling algorithm and fault tolerant mechanism, we have developed a simulation application based on SimGrid. With this simulation application, experiments are realized in several aspects.

7.7.1 Simulation setup

The simulation setup is shown in Figure 7.5. In the tree structure, we have one MMS, two LMSs and four SMSs and each scheduler is deployed in a server with 310 Mflop/s of CPU power. The bandwidth of connection between each scheduler is 10 Mbps. Moreover, each CR is a host which delivers 580 Mflops of CPU power and it can be considered as the head node of a cluster. The quantity of CPU in each cluster is marked below CR and the bandwidth of connection between CR and SMS is 100 Mbps. The client machine delivers 220 Mflops CPU power and connects MMS via Internet with 780 Kbs bandwidth.

Each application is assigned a number, e.g., 0, 1, 2, and they are deployed in CRs randomly. For example, in the site of SMS1, the first CR from the left is a cluster that has 6 CPUs and it supports the applications 0, 2 and 4. In the simulation, both MMS and SMS implement the minimum completion time (MCT) heuristic [Siegel and Ali, 2000] to make scheduling decisions.

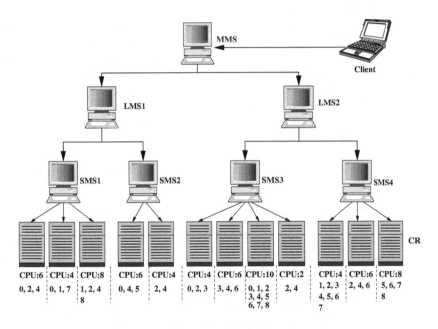

FIGURE 7.5: The simulation environment.

7.7.2 Comparison with centralized scheduling

The distributed scheduling structure can achieve a large-scale grid system. But comparing with centralized scheduling, this scheduling schema uses more message exchanges for a job submission and thus the job mapping in this schema should be less effective. In this experiment, a centralized scheduling schema is also simulated. A scheduler manages all the twelve CRs and submits directly jobs to these CRs.

Two clients are deployed and each submits 500 jobs to MMS with 2 seconds interval. Nine types of applications (0–9) are deployed and the client selects randomly the application type to submit. For each job, the client calculates its completion time which equals the result of subtracting its submitted time from its finished time. The submitted time is the time when the job is submitted by a client and the job's finished time is the time when the job is completed by a computing resource (CR). Then, for each type of application, the client calculates its average completion time. For the centralized scheduling schema, we launch the same experiment.

The result is shown in the Table 7.1 and Figure 7.6. From these results, we can conclude that the centralized scheduling and distributed scheduling give a similar average completion time, approximately 10% difference. But the difference between the average completion time and the estimated execution time varies vastly according to the application type. For example, the average completion time of the application 4 for the two scheduling schema (91.254651 and 94.80425) approaches closely the application's execution time (86.2069). However, for the application 8, the average completion time doubles its execution time. It is because the average completion time depends strongly on the application's deployment. There are nine CRs which support the application 4 and the application 8 is only deployed in three CRs. Thus, jobs which want to execute application 8 must wait a long time in a job queue to be executed.

Application type	$EET(s)^*$	$ACT(s)^*$ Centralized scheduling	$ACT(s)^*$ Distributed scheduling
0	8.620689655	8.955351	9.4926875
1	1.724137931	2.1749745	2.123462
2	17.24137931	23.3168755	22.074354
3	68.965517241	140.380316	169.3790235
4	86.206896552	91.254651	94.80425
5	172.413793103	515.6953335	556.0106805
6	120.689655172	208.4824085	225.087204
7	10.344827586	10.749057	10.751203
8	862.068965517	1743.946354	1951.589965
EET: Estimated Execution Time, ACT: Average Completion Time			

Table 7.1: The application estimated execution time and average completion time with two clients.

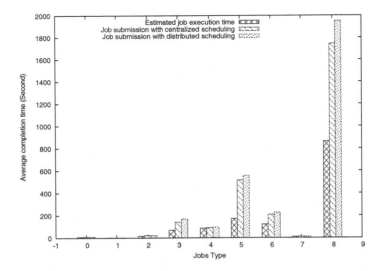

FIGURE 7.6: The average completion time for centralized and distributed scheduling with two clients.

For the centralized scheduling schema, another problem is the communication bottlenecks. When lots of clients send requests to the central scheduler in the same time, the scheduler will maintain the communication with each client and make a scheduling decision for each task in a very short duration that causes a heavy load in the host of scheduler and thus reduces the performance of the scheduling system. Our model has also the only entry (MMS) for clients; therefore it may have the communication bottleneck problem. But considering the tree structure of schedulers which can participate in the global scheduling work and make partial scheduling decisions, our model should be more effective in the case of multi-client's requests. Here a multi-client's request is defined as: each client in a multiple client's set sends a request to a scheduler in the same time.

In order to evaluate the performance of our model in the multi-client's requests case, another simulation is effectuated. In this simulation, 30 clients are deployed and each submits 20 jobs to MMS with 2 seconds interval. The client can randomly select an application type from the nine types of application (0–9) for the submission. Each client calculates the average completion time for every type of job when all its tasks have been completed. Then we calculate the average completion time for every type of job when all the clients have completed. For the centralized scheduling schema, we run the same experiment. The simulation results are shown in Table 7.2 and Figure 7.7.

According to the simulation results, we can conclude that the average com-

Application type	$EET(s)^*$	$ACT(s)^*$ Centralized scheduling	$ACT(s)^*$ Distributed scheduling
0	8.620689655	103.026531464	113.417606037
1	1.724137931	154.080089103	136.260599517
2	17.24137931	215.398029143	279.702666792
3	68.965517241	455.880322519	542.100991966
4	86.206896552	291.06166364	327.650004069
5	172.413793103	886.336876692	844.413221207
6	120.689655172	610.171255138	421.416273593
7	10.344827586	448.915829429	84.547881214
8	862.068965517	2154.134783423	1606.397786708
EET: Estimated Execution Time, ACT: Average Completion Time			

Table 7.2: The application estimated execution time and average completion time with thirty clients.

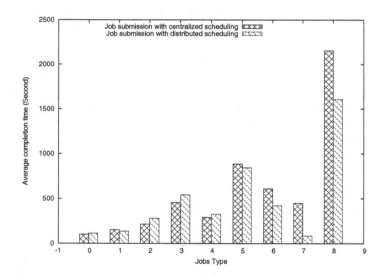

FIGURE 7.7: The average completion time for centralized and distributed scheduling with thirty clients.

pletion time for each type of job in this simulation is much longer than the simulation of two clients, because of the simultaneous application requests. As we have supposed, the distributed scheduling gives a more effective scheduling performance and the average completion time of most types of job is shorter than the time of central scheduling.

7.7.3 Fault tolerance experiments

In the model considered, the scheduling algorithm is fault tolerant and the failure can be detected and repaired automatically. Thus the performance of this algorithm and detection mechanism must be tested.

A client submits 100 jobs which execute application 5 with 2 seconds interval. Two experiments are effectuated: SMS4 failure after 70 seconds and LMS2 failure after 70 seconds. The last failure occurs after 100 seconds and then the failed component is recovered. In order to make a comparison, the client submits 100 jobs again without scheduler failure. The result is shown in Figure 7.8 and in Figure 7.9.

From these two figures, we can find that all the submitted jobs are completed though there is a failure of scheduler. In the case of SMS4 failure, all its CRs will detect the failure and connect to SMS3 to reconstruct the connection after near 30 seconds. During this treatment, more jobs are submitted to SMS2 although there are not more free CPU to execute the jobs and these jobs make a peak in the curve. In the case of LMS2 failure, SMS3 and SMS4 will detect their father's failure and connect to MMS after near 30 seconds. But in this period, both SMS3 and SMS4 are not available temporarily. Thus there are more jobs submitted to SMS2 than the case of SMS4 failure and in the curve of Figure 7.9, there are more jobs in the peak. For the recovery, the CRs or SMSs reconnect to their old father by manner of exchange messages and there is no more influence to jobs submitted in each CR.

7.7.4 Workload analysis

Load balancing is an important problem which is considered by most scheduling systems. Therefore, the performance of load balancing for our model is evaluated in this experiment. A client submits 500 jobs of application 4 to MMS with 2 seconds interval. The result in Table 7.3 illustrates that all the CRs which are deployed by application 4 have submitted jobs and the number of effectuated jobs in each CPU is similar. The completed jobs per CPU in the site of SMS2 are the biggest because they have the least number of CRs than the other three sites. Thus it takes less time to communicate with its sons and its responses are quicker.

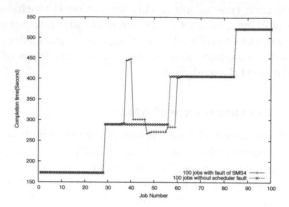

FIGURE 7.8: SMS failure experiment.

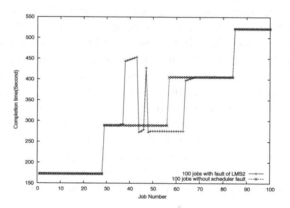

FIGURE 7.9: LMS failure experiment.

	Number of completed jobs (N)	CPU number for application 4 (C)	N/C
SMS1	143	14	10.21
SMS2	107	10	10.7
SMS3	154	18	8.56
SMS4	96	10	9.6
Total	500	52	9.62

Table 7.3: Statistics of submitted jobs of application 4 in each SMS.

7.8 Related work

In a general distributed system, scheduling algorithms can be classified into two types: static and dynamic. For static algorithms, the complete set of tasks to be scheduled is known a priori, the scheduling is done prior to the execution of any of the tasks and more time is available to make the scheduling decision. Nevertheless, dynamic methods perform the scheduling as tasks arrive. Dynamic scheduling is more appropriate than static scheduling in a grid environment because of the dynamic availability and load variability of computing resources [Siegel and Ali, 2000].

The dynamic scheduling algorithms can be grouped into two categories: online mode and batch-mode heuristics. In the batch mode, tasks are collected into a set that is prepared for scheduling at a predefined time interval. In contrast to the batch mode, the on-line mode maps a task onto a machine as soon as it arrives at the scheduler. The most adopted heuristics of online modes are minimum completion time (MCT), minimum execution time (MET) and KPB (k-percent best). According to the paper [Siegel and Ali, 2000], the KPB provides the minimum makespan, closely followed by the MCT. Moreover, sufferage heuristics are heuristics which give the smallest makespan in the batch mode heuristic.

In a cluster, single point of failure (SPoF) causes the cluster to go down completely. The paper [Limaye et al., 2005b] has proposed the HA-OSCAR solution for this type of failure. In this solution, a standby server which is a clone of a head node is deployed. A HA-OSCAR monitoring service is started in both the head node and its standby. This monitoring service monitors the status of critical services in the head node (e.g., resource service, Globus container, gridFTP in our model) and tries to restore these services when they are not working. If the monitoring service can not restart the failed service successfully after several attempts, the standby acquires the head node's public and private IP and thus becomes the head node. A smart failover/fail-back mechanism is implemented to periodically save job states in the job queue and synchronize those states to the standby server. When the standby server is called to action, it will then start jobs from its last saved job state. The HA-OSCAR solution solves well the cluster's head node failure problem, but it is a more expensive solution. For each head node of cluster, a standby server must be deployed to provide the uninterrupted serviceability.

7.9   Concluding remarks

This chapter presents a scalable and fault tolerant distributed scheduling model for application integration in a grid environment. In the proposed model, the fault tolerance issues are considered in two aspects: the scheduling algorithm level and the failure detection mechanism. The DDFT algorithm is a robust scheduling algorithm to ensure jobs submission and mapping even if there is a failure of scheduler or connection. Moreover a series of algorithms are proposed to detect the failed scheduler or connection and reconstruct automatically the scheduling structure. Thus the two aspects enhance the fault tolerance in the model and make a robust distributed scheduling system. The simulation evaluates that the scheduling performance of DDFT algorithm is near the results of centralized scheduling system and the detection and repairing of failure can be effectuated automatically and effectively.

The model does not take into account the bandwidth influence for the scheduling and fault tolerance. Since computing resources in the grid are normally connected by wide area network links (WAN), the bandwidth limitation is an issue that must be considered when running *data-intensive* applications on such environments. DDFT must be modified to take into account the connection parameter for mapping a job into a resource.

7.10 References

[Bayucan et al., 1999] Bayucan, A., Henderson, R., Lesiak, C., Mann, B., Proett, T., and Tweten, D. (1999). Portable batch system: external reference specification. Technical report, MRJ Technology Solutions.

[Berman et al., 2001] Berman, F., Chien, A., Cooper, K., Dongarra, J., Foster, I., Gannon, D., Johnsson, L., Kennedy, K., Kesselman, C., Mellor-Crumme, J., Reed, D., Torczon, L., and Wolski, R. (2001). The GrADS project: software support for high-level grid application development. *International Journal of High Performance Computing Applications*, 15(4):327–344.

[Bramley et al., 2000] Bramley, R., Chiu, K., Diwan, S., Gannon, D., Govindaraju, M., Mukhi, N., Temko, B., and Yechuri, M. (2000). A component based services architecture for building distributed applications. In *Proceedings of HPDC*, page 51.

[Brown, 2005] Brown, M. C. (2005). Build grid applications based on SOA. Technical report, MCslp.

[Buyya et al., 2002] Buyya, R., Abramson, D., and Giddy, J. (2002). A computational economy for grid computing and its implementation in the Nimrod-G resource broker. *Future Generation Computer Systems*, 18:1061–1074.

[Caron and Desprez, 2006] Caron, E. and Desprez, F. (2006). DIET: a scalable toolbox to build network enabled servers on the grid. *International Journal of High Performance Computing Applications*, 20(3):335–352.

[Casanova et al., 2008] Casanova, H., Legrand, A., and Quinson, M. (2008). SimGrid: a generic framework for large-scale distributed experimentations. In *Proceedings of the 10th IEEE International Conference on Computer Modelling and Simulation (UKSIM/EUROSIM'08)*. IEEE Computer Society.

[Christensen et al., 2001] Christensen, E., Curbera, F., Meredith, G., and Weerawarana, S. (2001). Available online at: `http://www.w3.org/TR/wsdl` (accessed May 1, 2009).

[Condor, 2009a] Condor (2009a). Condor-G. Available online at: `http://www.cs.wisc.edu/condor/manual/v6.4/5_3Condor_G.html` (accessed May 1, 2009).

[Condor, 2009b] Condor (2009b). User's manual. Available online at: `http://www.cs.wisc.edu/condor/manual/v6.8/2_4Road_map_Running.html` (accessed May 1, 2009).

[Dialani et al., 2002] Dialani, V., Miles, S., Moreau, L., Roure, D. D., and Luck, M. (2002). Transparent fault tolerance for web services based architectures. In *Proceedings of the 8th International Euro-Par Conference on Parallel Processing (Euro-Par'02)*, pages 889–898, London, UK. Springer-Verlag.

[EGEE, 2005] EGEE (2005). Egee middleware architecture and planning (release 2). Technical report, DJRA1.4, EGEE.

[EGEE, 2009] EGEE (2009). gLite-ligthweight middleware for grid computing. Available online at: `http://glite.web.cern.ch/glite/` (accessed May 1, 2009).

[Fallside, 2001] Fallside, D. (2001). Available online at: `http://www.w3.org/TR/xmlschema-0` (accessed May 1, 2009).

[Foster, 2005] Foster, I. (2005). Globus toolkit version 4: software for service-oriented systems. In *Proceedings of the International Conference on Network and Parallel Computing (IFIP)*, volume 3779 of *Lecture Notes in Computer Sciences*, pages 2–13. Springer-Verlag.

[Foster et al., 2005] Foster, I., Czajkowski, K., Ferguson, D., Frey, J., Graham, S., Maguire, T., Snelling, D., and Tuecke, S. (2005). Modeling and managing state in distributed systems: the role of OGSI and WSRF. In *Proceedings of the IEEE Conference*, volume 3, pages 604–612. IEEE Computer Society.

[Foster et al., 2002a] Foster, I., Kesselman, C., Nick, J., and Tuecke, S. (2002a). Grid services for distributed system integration. *IEEE Computer Society*, 35:37–46.

[Foster et al., 2002b] Foster, I., Kesselman, C., Nick, J., and Tuecke, S. (2002b). The physiology of the grid: an open grid services architecture for distributed systems integration. In *Open Grid Service Infrastructure WG, Global Grid Forum*. Available online at: `citeseer.ist.psu.edu/foster02physiology.html` (accessed May 1, 2009).

[Foster et al., 2002c] Foster, I., Kesselman, C., Nick, J. M., and Tuecke, S. (2002c). Grid services for distributed system integration. *Computer*, 35(6):37–46.

[Foster et al., 2001] Foster, I., Kesselman, C., and Tuecke, S. (2001). The anatomy of the grid: enabling scalable virtual organizations. *International Journal of High Performance Computing Applications*, 15(3):200–222.

[Frey et al., 2002] Frey, J., Tannenbaum, T., Livny, M., Foster, I., and Tuecke, S. (2002). Condor-G: a computation management agent for multi-institutional grids. *Cluster Computing*, 5(3):237–246.

[Gannon et al., 2003] Gannon, D., Ananthakrishnan, R., Krishnan, S., Govindaraju, M., Ramakrishnan, L., and Slominski, A. (2003). Grid web services and application factories. *Computing.*

[Globus, 2009a] Globus (2009a). globusrun-WS: official job submission client for WS-GRAM. Available online at: `http://www.globus.org/toolkit/docs/4.0/execution/wsgram/rn01re01.html` (accessed May 1, 2009).

[Globus, 2009b] Globus (2009b). GT4.0. Available online at: `http://www.globus.org` (accessed May 1, 2009).

[Globus, 2009c] Globus (2009c). GT4.0 WS-GRAM: job description schema documentation. Available online at: `http://www.globus.org/toolkit/docs/4.0/execution/wsgram/schemas/gram_job_description.html` (accessed May 1, 2009).

[Globus, 2009d] Globus (2009d). Submitting a job in Java using WS-GRAM. Available online at: `http://www.globus.org/toolkit/docs/4.0/execution/wsgram/WS_GRAM_Java_Scenarios.html` (accessed May 1, 2009).

[Gracjan et al., 2006] Gracjan, J., Radoslaw, J., Rafal, M., and Jozsef, K. (2006). Grid checkpointing architecture: a revised proposal. Technical Report TR-0036, Institute on Grid Information, Resource and Workflow Monitoring Systems, CoreGRID - Network of Excellence. Available online at: `http://www.coregrid.net/mambo/images/stories/TechnicalReports/tr-0036.pdf` (accessed May 1, 2009).

[GridLab, 2004] GridLab (2004). Grid application toolkit. Available online at: `http://www.gridlab.org/WorkPackages/wp-1` (accessed May 1, 2009).

[Gridlab, 2005] Gridlab (2005). Products and technologies. Available online at: `http://www.gridlab.org/about.html` (accessed May 1, 2009).

[Hamscher et al., 2000] Hamscher, V., Schwiegelshohn, U., Streit, A., and Yahyapour, R. (2000). Evaluation of job-scheduling strategies for grid computing. In *Proceedings of the 1st IEEE/ACM International Workshop on Grid Computing (GRID'2000)*, Lecture Notes in Computer Sciences, Berlin, Heidelberg. Springer-Verlag.

[Huang et al., 2003] Huang, Y., Taylor, I., Walker, D., and Davies, R. (2003). Wrapping legacy codes for grid-based applications. In *Proceedings of the International Parallel and Distributed Processing Symposium.*

[Huedo et al., 2007] Huedo, E., Montero, R. S., and Llorente, I. M. (2007). A modular meta-scheduling architecture for interfacing with pre-WS and WS-grid resource management services. *Future Generation Computer Systems*, 23:252–261.

[Imamagic et al., 2006] Imamagic, E., Radic, B., and Dobrenic, D. (2006). An approach to grid scheduling by using Condor-G matchmaking mechanism. In *Information Technology Interfaces*, pages 625–632.

[Kacsuk et al., 2004] Kacsuk, P., Goyeneche, A., Delaitre, T., Kiss, T., Farkas, Z., and Boczko, T. (2004). High-level grid application environment to use legacy codes as OGSA grid services. In *Proceedings of the 5th IEEE/ACM International Workshop on Grid Computing (GRID'04)*, pages 428–435, Washington, DC, USA. IEEE Computer Society.

[Kandaswamy et al., 2005] Kandaswamy, G., Fang, L., Huang, Y., Shirasuna, S., and Gannon, D. (2005). A generic framework for building services and scientific workflows for the grid. In *Proceedings of the 2005 ACM/IEEE Conference on SuperComputing*. IEEE Computer Society.

[Krishnan et al., 2001] Krishnan, S., Bramley, R., Govindaraju, M., Indurkar, R., Slominski, A., Gannon, D., Alameda, J., and Alkaire, D. (2001). The XCAT science portal. In *Proceedings of SC2001*, pages 49–49, New York, NY, USA. ACM Press.

[Kuebler and Eibach, 2002] Kuebler, D. and Eibach, W. (2002). Adapting legacy applications as web services. IBM DeveloperWorks. Available online at: http://www-128.ibm.com/developerworks/library/ws-legacy/ (accessed May 1, 2009).

[Letondal, 2004] Letondal, C. (2004). PISE: a tool to generate web interfaces for molecular biology programs. Available online at: http://www.pasteur.fr/recherche/unites/sis/Pise (accessed May 1, 2009).

[Limaye et al., 2005a] Limaye, K., Leangsuksun, B., Greenwood, Z., Scott, S., Engelmann, C., Libby, R., and Chanchio, K. (2005a). Job-site level fault tolerance for cluster and grid environments. In *Proceedings of the IEEE International Cluster Computing*, pages 1–9, Burlington, MA. IEEE Computer Society.

[Limaye et al., 2005b] Limaye, K., Leangsuksun, B., Munganuru, V. K., Greenwood, Z., Scott, S. L., Libby, R., and Chanchio, K. (2005b). Grid-aware HA-OSCAR. In *Proceedings of the 19th International Symposium on High Performance Computing Systems and Applications (HPCS'05)*, pages 333–339, Washington, DC, USA. IEEE Computer Society.

[Limaye et al., 2005c] Limaye, K., Leangsuksun, C., and Tikotekar, A. (2005c). Fault tolerance-enabled HPC scheduling with HA-OSCAR framework. In *Proceedings of High Availability and Performance Workshop (HAPCW'2005)*, Santa Fe, NM, USA.

[Lodygensky et al., 2003] Lodygensky, O., Fedak, G., Cappello, F., Neri, V., Livny, M., and Thain, D. (2003). XtremWeb & Condor : sharing resources between internet connected Condor pool. In *Proceedings of the*

3rd IEEE/ACM International Symposium on the Cluster Computing and the Grid (CCGrid 2003), pages 382–389. IEEE Computer Society.

[Mausolf, 2005] Mausolf, J. (2005). Grid in action: monitor and discover grid services in an SOA/web services environment. Available online at: `http://www-128.ibm.com/developerworks/grid/library/gr-gt4mds/index.html` (accessed May 1, 2009).

[Platform, 2003] Platform (2003). Open source metascheduling for virtual organizations with the community scheduler framework (CSF). Technical report, Platform Computing.

[Qi et al., 2006] Qi, L., Jin, H., Foster, I., and Gawor, J. (2006). HAND: highly available dynamic deployment infrastructure for globus toolkit 4.

[Schopf et al., 2005] Schopf, J. M., D'Arcy, M., Miller, N., Pearlman, L., Foster, I., and Kesselman, C. (2005). Monitoring and discovery in a web services framework: functionality and performance of the globus toolkit's mds4. Technical report, Preprint ANL/MCS-P1248-0405, Argonne National Laboratory, Argonne, IL.

[Seidel et al., 2002] Seidel, E., Allen, G., Merzky, A., and Nabrzyski, J. (2002). GridLab: a grid application toolkit and testbed. *Future Generation Computer Systems*, 18(8):1143–1153.

[Siegel and Ali, 2000] Siegel, H. J. and Ali, S. (2000). Techniques for mapping tasks to machines in heterogeneous computing systems. *Journal of Systems Architecture*, 46:627–639.

[Silva, 2006] Silva, V. (2006). Quick start to a GT4 remote execution client. Available online at: `http://www-128.ibm.com/developerworks/grid/library/gr-wsgram/` (accessed May 1, 2009).

[Sotomayor, 2009] Sotomayor, B. (2009). The globus toolkit 4 programmer's tutorial.

[Sundaram, 2005a] Sundaram, B. (2005a). Introducing GT4 security. Available online at: `http://www-128.ibm.com/developerworks/grid/library/gr-gsi4intro/` (accessed May 1, 2009).

[Sundaram, 2005b] Sundaram, B. (2005b). WS-notification and the globus toolkit 4 WS-Java core. Available online at: `http://www-128.ibm.com/developerworks/grid/library/gr-wsngt4/` (accessed May 1, 2009).

[Thain et al., 2005] Thain, D., Tannenbaum, T., and Livny, M. (2005). Distributed computing in practice: the Condor experience. *Concurrency: Practice and Experience*, 17(2–4):323–356.

[Tonellotto et al., 2005] Tonellotto, N., Wieder, P., and Yahyapour, R. (2005). A proposal for a generic grid scheduling architecture. In *Proceedings of the Integrated Research in Grid Computing Workshop*, pages 20–28, Pisa, Italy.

[W3C, 1999] W3C (1999). XML path language (XPath) version 1.0. Available online at: http://www.w3.org/TR/xpath (accessed May 1, 2009).

[W3C, 2009] W3C (2009). Web services description language (WSDL) 1.1. Available online at: http://www.w3.org/TR/wsdl (accessed May 1, 2009).

[Yu and Magoulès, 2007] Yu, L. and Magoulès, F. (2007). A framework for dynamic deployment of scientific applications based on wsrf. In *Advances in Grid and Pervasive Computing*, Paris, France. Springer-Verlag.

Chapter 8

Broadcasting for grids

Christophe Cérin

LIPN, Université de Paris Nord, 99 avenue J.B. Clément, 93430 Villetaneuse, France

Luiz-Angelo Steffenel

CReSTIC, Université de Reims Champagne-Ardenne, BP 1039, 51687 Reims Cedex 2, France

Hazem Fkaier

Unité de recherche UTIC / ESSTT, 5 avenue Taha Hussein, B.P. 56, Bab Mnara, Tunis, Tunisia & LIPN, Université de Paris-Nord, 99 avenue J.B Clément, 93430 Villetaneuse, France

8.1 Introduction

It is well known that communication affects significantly the performance of applications deployed on large scale architectures. Large scale platforms are characterized by a collection of a great number of computing resources that are geographically distributed over some sites and connected with a wide heterogeneous dedicated network. Since data sizes in Grid applications may be large as well as the number of nodes, the collective communication inherent to the applications is a critical bottleneck.

In this chapter we consider institutional grids. We mean clusters of clusters in geographically different places but running in the same VLAN. The underlying use case is familiar for the scientist who develops parallel codes, say with the MPI language. After the development step, he reserves nodes in the grids, as many as he wants and maybe with the Kadeploy[1] tool, and he does not take care about the node's location. His MPI application contains collective communication primitives and he also does not want to tailor his application to get performance according to the node's choice. As pointed out in an article about profiling parallel programs [Rabenseifner, 1999], Rabenseifner justifies his work on the AllReduce primitive because 8.54% of the total execution time in applications is due to collective operations and among this 8.54% of execution time 37% is for the `MPI_Allreduce`. Rabenseifner does not provide information about the `MPI_Bcast` (one to all) but draws charts about the send, isend, ireceive, reduce primitives.

In this context we introduce broadcasting algorithms for the grids corresponding to the `MPI_Bcast` operation. In the long term, we are focusing on efficient parallel programming libraries for the grid able to run efficiently on heterogeneous platforms. For this problem, we need to investigate the issues of communication cost between clusters and inside a cluster and of course how to schedule communication. Since one of our goals is to implement algorithms and to validate them with experiments, we introduce in this chapter some experimental results.

We do not consider here broadcasting securely. We mean that we do not consider any fault-tolerant broadcasting schema nor inter-communicator collective operations, that is to say operations that send data from one member of one group to all members of the other group. Such refinement of the problem allows optimizations in the message order but it is quite difficult to solve without a strong background. Here we consider only one group. Please refer to [Jelena et al., 2005] for an implicit work on performance analysis of MPI collective operations.

The organization of the chapter is the following. In Section 8.2, we recall some definition and theoretical results. In Section 8.3 we introduce techniques based on heuristics. In Section 8.4 we introduce related methods such as dynamic programming approach and a multi-criteria approach. Section 8.5 concludes the chapter.

8.2 Broadcastings

From a theoretical point of view, the problem of broadcasting may find its root in the construction of partial minimum spanning trees which is a problem in graph theory. A graph depicts the underlying infrastructure: nodes in the graph represent the machines and edges represent the cost of sending a message, say between two machines.

---

[1]See `https://gforge.inria.fr/projects/kadeploy/`.

A graph often contains redundancy in that there can be multiple paths between two vertices. This redundancy may be desirable, for example to offer alternative routes in the case of breakdown or overloading of an edge (road, connection, phone line) in a network. However, we often require the cheapest sub-network that connects the vertices of a given graph. This must in fact be an unrooted tree, because there is only one path between any two vertices in a tree; if there is a cycle then at least one edge can be removed. The total cost or weight of a tree is the sum of the weights of the edges in the tree. We assume that the weight of every edge is greater than zero. Given a connected, undirected graph $G =< V, E >$, the minimum spanning tree problem is to find a tree $T =< V, E' >$ such that $E' \subseteq E$ and the cost of T is minimal.

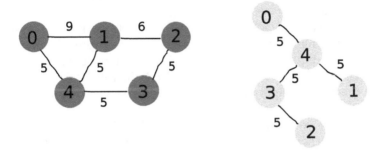

FIGURE 8.1: A graph with 5 vertices (left) and its minimum spanning tree (right).

Figure 8.1 is an example of a graph and its minimum spanning tree. Now, for broadcasting a message from vertices 0 to all the others, we can send from vertices 0 to vertices 4, then from vertices 4 to vertices 3, then vertices 4 can redistribute to vertices 1 and finally vertices 3 sends the message to vertices 2. Computing a solution of this problem can be done with Prim's or Kruskal's Algorithms at a cost time of $\mathcal{O}(|V|^2)$ and $\mathcal{O}(|V| \log |V|)$ respectively.

Notice that in the previous example, in order to optimize the total execution time we allow vertices 4 to redistribute the message after a first send. In a 2-port model, we guess that we can send simultaneously on the 2 ports, that is to say to vertices 3 and 1 at the same time. The total execution time is the length of the critical path in the tree.

The correct formulation of the problem is related to the minimum broadcast time[2] where the problem is stated as follows:

INSTANCE: Graph $G = (V, E)$ and a source node $v_0 \in V$.

---

[2]See: `http://www.nada.kth.se/\~{}viggo/wwwcompendium/node127.html\#5671`.

SOLUTION: A broadcasting scheme. At time 0 only v_0 contains the message that is to be broadcast to every vertex. At each time step any vertex that has received the message is allowed to communicate the message to at most one of its neighbors.

MEASURE: The broadcast time, i.e., the time when all vertices have received the message.

This has been termed the minimum broadcast time problem under the telephone model and is known to be NP-complete. The minimum broadcast time in a graph has a solution that is approximable within $O(\log^2 |V| / \log \log |V|)$ [Ravi, 1994]. Instead of implementing the result, we go along heuristics. We have no idea, to our knowledge, if the theoretical result has been implemented for large scale grid systems.

We shall also emphasize here that optimal broadcast tree is fundamentally different from minimal spanning tree (MST). In the optimal broadcast tree the issue is to minimize the time to reach the last node which is to say to minimize the longest path in the tree. While in the MST, the issue is to minimize the whole weight of the tree. These two constructions may lead to very different trees.

In this chapter, we study first heuristics for the minimum broadcasting tree and we evaluate them through simulations according to parameters extracted from the Grid'5000[3] testbed. Grid'5000 is a Grid testbed composed of 5000 processors distributed over some clusters in 9 sites in France. The inter-cluster connection is insured by RENATER Education and Research Network. All clusters are connected to Renater with a 10Gb/s link.

The fact that all clusters are linked through an homogeneous network with the same high bandwidth (the network links are all optic fiber based) means that we can suppose that sending a message of given size would take the same lap of time for any couple sender-receiver independently of geographic distances.

According to our hypothesis about the architecture, we have some supplementary difficulties with implementations since the network is heterogeneous and, in our case, structured in two levels: inter-cluster level and intra-cluster level. Contentions may occur and special techniques have been elaborated in [Steffenel, 2006] for this case. Many heuristics have been imagined to approximate the best way to broadcast a message in a cluster and in a cluster of clusters [Barnett et al., 1996], [Matsuda et al., 2004], [Matsuda et al., 2006], [Bhat et al., 2003], [Barchet-Steffenel and Mounie, 2006], [Steffenel and Mounié, 2008]. We propose in this chapter to review them and to combine them in order to exhibit a new one that behaves as well as the better known heuristics.

---

[3]See https://www.grid5000.org.

Second, in current cluster of clusters architecture, the number of clusters is reduced. Grid'5000 for instance is composed of 9 sites. Hence examining all possible broadcast tree is not a so hard computational task. Dynamic programming may be used in this case. We follow the approach introduced in [Park et al., 1996] for one homogeneous cluster. We propose a generalization of this approach to the case of cluster of clusters.

Third we will summarize other work in the direction of mastering communication for collective operations.

8.3 Heuristics for broadcasting

One of the earliest papers, to our knowledge, dealing with communication in a wide sense and in a grid is the paper of Ian Foster [Foster and Karonis, 1998]. In this paper authors investigated the need for a communication library in the frame of parallel programming on a distributed multi-site architecture where heterogeneity of network is an intrinsic property. They proposed a version of MPICH dedicated to grids and they called it MPICH-G. This version is built upon MPICH and Globus.

Other studies have been elaborated. Most of them consider the MPI communication library or one of its variants such as MPICH-G2, PACX-MPI, GridMPI. To deal specifically with broadcasting, we need to refer to the algorithm of Van de Geijn [Barnett et al., 1996] which consists in a recursive scatter (a special broadcast) phase that puts a fragment of the message to distribute on each node; then a phase of recursive all-gather (a concatenation of messages that are stored on each node) occurs to each message fragment. For the sake of completeness, the algorithm consists of one-to-many messages (authors call them scatter) and steps of many-to-many messages (authors call them gather).

We can also cite the works achieved in "The Grid Technology Research Center, AIST" by M. Matsuda et al. In [Matsuda et al., 2006 , Matsuda et al., 2004], Matsuda et al. studied collective operations on multi-site architectures with MPI library. The paper [Tatsuhiro et al., 2004] of Satochi et al. considers especially the broadcast operation in the case where nodes have 2 lane network interface controllers. The main contribution of the paper is the way of splitting the message to broadcast: it is broken into two pieces and then they are broadcasted independently following two binary trees; then, nodes of the two trees exchange the two parts of the message.

Let us review now more in detail some of the known algorithms for both homogeneous and heterogeneous environments.

8.3.1 Basic approaches for broadcasting in homogeneous environments

We review hereby, some well-known algorithms for broadcasting a message in one cluster. In the remainder, we assume that we can reach any node in an equal time. Then, there is no need to choose a specific node since network is homogeneous and all nodes are symmetric.

Linear broadcast: It is the simplest basic algorithm. It consists in putting all nodes in a queue where the node that will initiate the broadcast is the head. The broadcast algorithm begins when the head sends the message to its following in the queue, then progressively the message is sent from one to his following until it reaches the tail node. We can also proceed with one level (flat) tree where the root is the node detaining initially the message and all the other nodes are leaves. Then the root sends sequentially the message to the leaves. The broadcast time in both case is proportional to the processor's number.

Binary tree algorithm: We can improve the above algorithm by proceeding with a binary tree. The node detaining the message to be broadcasted (let's call it P_0) sends it to two other nodes (we call them P_1 and P_2). In the second step, P_0 stays idle, and only P_1 and P_2 send each to two other nodes ($P_3 \ldots P_6$). In the next step P_1 and P_2 stay idle, and $P_3 \ldots P_6$ send each to two other nodes, and so on. According to the network model based on (l, b, M) parameters to estimate the cost of broadcasting, where l is the network latency, b is network bandwidth and M is message size, we derive easily the broadcast cost is $2(l + M/b)\log p$, since the height of a binary tree of p nodes is $\log p$ and each node makes two sends.

Binomial tree algorithm: We can yet improve the above algorithm by letting all processors having the message at a given time participate in the following broadcast steps. Then we do not need any more than one node send the message two times per step. The algorithm proceeds as follows: in the first step, P_0 sends to P_1. In the next step, both P_0 and P_1 send to two other nodes (P_2 and P_3). Then all processors detaining the message send it to four other processors and so on.

The cost of broadcasting according to such tree is $(l + M/b)logp$, since the number of nodes detaining the message doubles at each step.

Following another approach, the Van de Geijn algorithm [Barnett et al., 1996] proposes to act in two steps: a recursive scatter in a binomial tree fashion, then collecting pieces using recursive doubling until the whole message becomes available on all nodes. The complexity of this algorithm is proportional to the size of the message ($O(M)$) since the message is split into two pieces independently following two binary trees. We do not detail this algorithm because we do not use it in the remainder of the chapter.

8.3.2 Advanced approaches for heterogeneous clusters

We assume now that the network is heterogeneous in one cluster. Then, we shall wonder at each time which is the next destination of the message and which node detaining the message is to be chosen to achieve the remaining sending steps. We imagine that nodes are separated into two sets. The set A is the set of node already having the message at a given time. Set A is also called the set of candidate nodes to achieve the next send operation. The set B is the set of nodes that have not yet received the message. Set B is also called the set of candidate nodes to receive the message in the next communication step.

Initially, set A contains only the node that will initiate the broadcast operation. Many heuristics [Bhat et al., 2003], [Barchet-Steffenel and Mounie, 2006], [Steffenel and Mounié, 2008] have been elaborated to achieve the whole broadcast operation in optimal time. It is obvious that optimality is achieved not only by using all communication potential, but also by suitably fastest communication links. It is also known that identifying the best broadcasting tree is an NP-complete problem.

We introduce now some known heuristics that try to approximate the optimal broadcast tree. Note that optimal broadcast tree is not the minimal covering tree because the latter minimizes the cost of edges but does not take in account the maximization of communication potentials (nodes communicate in parallel).

Early Completion Edge First - ECEF: According to the ECEF heuristic (Bhat et al. [Bhat et al., 2003]) we choose a couple of nodes, P_i in set A and P_j in set B. In the next step, the message is sent from P_i to P_j. The couple (P_i, P_j) is chosen in such a way that P_j becomes ready to send the message as early as possible. This time is computed by:

$$RT_i + g_{ij}(m) + L_{ij}$$

where RT_i is the ready time of P_i, $g_{ij}(m)$ is the latency gap between P_i and P_j and L_{ij} is the communication cost between P_i and P_j. Note that this heuristic aims at increasing the number of nodes in set A as fast as possible.

Early Completion Edge First with look-ahead - ECEF-LA: It is clear that the ECEF heuristic allows increasing the number of nodes in set A which is yet a good fact. But it also important to well choose the next destination to be itself a good sender in the remaining steps.

As an enhancement of the latter heuristic, Bhat et al. [Bhat et al., 2003] propose to estimate the efficiency of each node throughout a function that takes into consideration the speed of forwarding the message to another node of set B. For instance the following function can be considered:

$$F_j = min(g_{jk}(m) + L_{ik})$$

for P_k in set B. Then we select in set B the node P_j that minimizes

$$RT_i + g_{ij}(m) + L_{ij} + F_j.$$

From a complexity point of view, ECEF has a running time of $\mathcal{O}(N^2 \log N)$, whereas, due to the look-ahead function, ECEF-LA has a running time of $\mathcal{O}(N^4)$.

8.3.3 Grid aware heuristics

We suppose, now, that we have a cluster of clusters environment. We suppose also that we have a coordinator (proxy) on each cluster. All communication between clusters are insured by these coordinators. Subsequently, global communications are ordered in two levels: inter-cluster communication and intra-cluster communication. Hence, if we have a message to broadcast through a grid architecture, then it is broadcasted between coordinators, and then each coordinator broadcasts the message locally. In the works elaborated by Steffenel [Steffenel and Mounié, 2008], authors simulate the local communication load by only one virtual node that is connected to a specific coordinator. Then the local communication load is depicted by

$$L_{kk'} + g_{kk'} = \begin{cases} T_k & \text{if } k' \text{ is associated to node } k \\ \infty & \text{if } k' \text{ is not associated to node } k \end{cases} \qquad (8.1)$$

where P_k is a coordinator and $P_{k'}$ is a virtual node simulating a cluster.

Under this framework, Steffenel proposed three heuristics to broadcast a message in a grid environment. The main question we are facing is: which is the next coordinator (cluster) to contact while keeping in mind the global communication load (between clusters) and local communication load of each coordinator. Let us review the heuristics.

ECEF-LA*t*: The first heuristic proposed in this context is the one that increases the number of nodes in set A in least time. Then, we choose at each time the coordinator that takes the least time to join set A as in ECEF-LA heuristic. It also adopts the look-ahead option. The efficiency function F_j is set to:

$$F_j = min(g_{ij}(m) + L_{jk} + T_k) \text{ for } k \in \text{ set B.}$$

ECEF-LA*T*: The previous heuristic encourages the coordinator with the lowest load at each step, which may imply delays on the most loaded coordinators and subsequently it may increase the broadcast completion time. The opposed heuristic is to choose at each time the coordinator that have the greatest load, i.e., the one that maximizes F_j. Hence, F_j is set to:

$$F_j = max(g_{ij}(m) + L_{jk} + T_k) \text{ for } k \in \text{ set B.}$$

BottomUp: It is clear that the last heuristic can not be optimal because we choose at each step the least powerful coordinator; so the number of nodes in set A increases very slowly. The last proposed heuristic in [Steffenel and Mounié, 2008] combines ECEF-LA*t* and ECEF-LA*T*. We need to begin by contacting the most loaded coordinator. We also need to contact it through the 'shortest path.' Then BottomUp heuristic uses a min-max approach to find the 'shortest path' to contact the most loaded coordinator. Hence it selects the coordinator verifying:

$$max_{P_j \in B}(min_{P_i \in A}(g_{ij}(m) + L_{ij} + T_j))$$

8.3.4 New approach for broadcasting in clusters and hyper clusters

According to previous heuristics, to reduce global broadcast time, three factors impact performance. First, we need to align set A with clusters, in the quickest way possible. Having more potential senders gives more of a chance to perform the next communication in a better way, since we have more choices to consider.

The second factor is to give an advantage to communication-efficient cluster when choosing a receiver. As we explained before, it is important to communicate with the cluster that can forward the message, in the next steps, within a short time. This means that we want to augment set A with good senders.

The third factor is to begin by contacting the most loaded clusters, so that we insure the maximum of overlap between intra and inter-cluster broadcast. This strategy is the key of success of BottomUp and ECEF-LAT heuristics since, according to measured parameters, local broadcast needs more time than inter-cluster communication.

The problem of building the optimal broadcast tree is not a multi-criteria problem, since we have only one objective function which is the minimization of the global broadcasting time. However, we believe that it is possible to transform each factor to an independent criteria that we have to 'optimize.' Then we can try to solve the whole problem as a multi-criteria problem.

New heuristic

The previous heuristics try to optimize one of these 'criteria' or to combine two or all of them at each iteration. And the better we consider these factors, the better the result is.

Each heuristic has a function to minimize. This function contains parameters linked to one or two of aforementioned factors. Hence all factors are merged in only one formula to minimize at each iteration.

Merging all factors in one may give us a compromised solution. But compromise is not always a good solution. To explain this idea let us imagine the situation where we have, at a given iteration, a 'very' loaded cluster. Then we should contact it in priority otherwise it will delay the ending time.

If we combine all factors and look for a compromise, then previous heuristics may lead us to choose another cluster, not the most loaded, and subsequently we do not achieve the best performance.

The same reasoning can be applied if we have a very good forwarder cluster or a very fast-to-communicate cluster at a given iteration. The conclusion of this example is to say that considering a single factor at a time can also be very efficient and even more efficient than combining much factors in one.

Following this idea we developed a new heuristic that considers each factor in a separated way. We proceed as follows:

We consider our two sets A and B. At each iteration we choose one sender from set A and one receiver from set B. Then at each iteration we shall decide which factor we need to satisfy: either (1) to choose the fastest-to-communicate cluster from set B, or (2) best forwarder cluster from set B or (3) the most loaded cluster form set B.

Condition (1) implies to minimize $RT_i + g_{ij}(m) + L_{ij}$ which is to say to apply one iteration of ECEF heuristic.

Condition (2) implies to minimize $RT_i + g_{ij}(m) + L_{ij} + F_j$ which is to say to apply one iteration of ECEF-LA heuristic.

And condition (3) implies to choose the most loaded cluster in set B and then to find the best sender in set A which is to say to apply one iteration of BottomUp heuristic, i.e., we apply $max_{P_j \in B}(min_{P_i \in A}(g_{ij}(m) + L_{ij} + T_j))$.

The question now is "How to choose the factor to satisfy?"

To answer this question, let us see what would happen if we do not satisfy a factor, i.e., we do not choose the best cluster according to this factor:

a- Either the chosen cluster (the optimal cluster according to another factor) behaves well with the considered factor then the considered factor is not strongly violated. Then we estimate that both clusters behave in relatively similar way according to that factor.

b- Or the chosen cluster behaves badly with the designated factor and then it violates the designated factor. Then we estimate that the clusters are relatively different for that factor.

At the end, it is important to choose the cluster that satisfies one factor and behaves well with the others or at least does not violate them strongly.

We propose to compute the set of values associated to each factor as follows: For factor (1), we compute set E1= $min_{P_i \in A}(RT_i + g_{ij}(m) + L_{ij}) / P_j \in B$. For factor (2), we compute set E2= $min_{P_i \in A}(RT_i + g_{ij}(m) + L_{ij} + F_j) / P_j \in B$. For factor (3), we compute set E3= $min_{P_i \in A}(RT_i + g_{ij}(m) + L_{ij} + T_j) / P_j \in B$.

Having dispersed value in a given set means that clusters are very different according to the associated factor; then the factor may be strongly violated if we do not satisfy it. Whereas, if a set contains close values, then it means that clusters behave in a quite similar way. Subsequently, choosing one cluster or an other will not be decisive.

Finally we choose to satisfy the factor which has the associated set with the most dispersed values, i.e., we compute the mean deviation of each set values and we choose to satisfy the factor having the greatest mean deviation.

Simulation

In our simulations, we rely on works done by Steffenel [Steffenel and Mounié, 2008] for the different parameters measured on a real grid environment. He measured values of different communications parameters (L, g, T) over the French grid infrastructure named Grid'5000.[4] He found out a lowest value and highest value for each parameter. In his simulation, he set randomly the values of L, g, T in the corresponding interval and then he applied the different heuristics. In our simulation we do the same. The values that we introduced now are the mean of 100 iterations.

Parameter	Min value	Max value
L	1	15
g	100	600
T	200	3000

Table 8.1: Grid'5000 settings (1/2).

In the first simulation, we set L, g and T in intervals measured over Grid'5000; see Table 8.1. As seen in Figure 8.2, all heuristics give almost the same completion time. Then we can not evaluate the efficiency or compare them. The second remark we shall note is that our new heuristic (noted MostCrit) gives exactly the same values as BottomUp which is the best heuristic at present time. We can conclude that both heuristics behave exactly in the same way and it can be obtained only if our new heuristic chooses to apply BottomUp at each iteration. By observing parameters values we can expect this fact since the interval of T_j is much larger than intervals of L_{ij} and g_{ij}. This means that values of T_j will be more sparse than values of L_{ij} and g_{ij} and consequently values in E3 will be more sparse than those in E1 and those in E2. And finally factor(3) (choosing the most loaded cluster) will be retained.

To evaluate the efficiency of our new heuristics, we proposed to achieve simulations with other settings. We changed the ratio of L and g (parameters

---

[4]For details, refer to `https://www.grid5000.fr`.

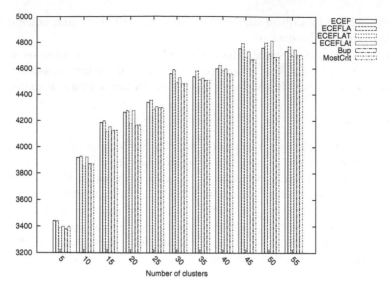

FIGURE 8.2: Broadcasting time versus clusters number with Grid'5000 settings.

linked to inter-cluster broadcast) and T (parameter associated to the local broadcast).

In the second simulation, top part of Figure 8.3, we multiplied L and g by 5 and divided T by 5; see top part of Table 8.2. In the third simulation, bottom part of Figure 8.3, we multiplied L and g by 10 and divided T by 10; see bottom part of Table 8.2.

	Top part of Figure 8.3		Bottom part of Figure 8.3	
Parameter	Min value	Max value	Min value	Max value
L	5	75	10	150
g	500	3000	1000	6000
T	40	600	20	300

Table 8.2: Grid'5000 settings (2/2).

Simulation represented in Figure 8.2 shows that BottomUp keeps giving good performances as well as our new heuristic 'MostCrit' even though they do not give exactly the same values. Other heuristics behave worse.

The final conclusion of our simulations is that BottomUp and MostCrit heuristics give evenly good results independently of the ratio of inter-cluster

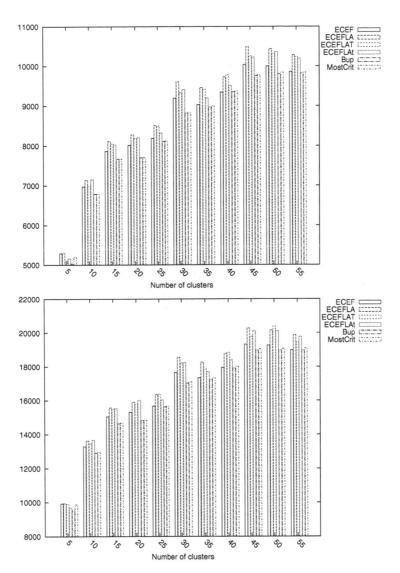

FIGURE 8.3: Broadcasting time versus clusters number with other setting.

communication performances over intra-cluster communication performances. This point has never been observed before to our knowledge.

8.4 Related work and related methods

8.4.1 Broadcasting and dynamic programming

In this section we give an exact solution for the problem of broadcasting one message in a cluster of clusters under some assumptions that we introduce now. The starting point is the work done by Lionel Ni in [Park et al., 1996] and it is related to dynamic programming for solving the equations defining the problem. Our target architecture is still a cluster of clusters whereas L. Ni work considers only a single cluster.

Point to point communication

In this paragraph, we recall the analysis of broadcasting as introduced in [Park et al., 1996]. First, the authors consider t_{send} which corresponds to the sending software latency. It includes the overhead of 'packetization,' checksum computing and possibly memory copying. Then they consider t_{recv} which corresponds to the receiving software latency. Its definition is similar to t_{send}. At last, the authors also consider t_{net} which is the time needed by a message to cross the network. It includes network parameters such as switching mechanism.

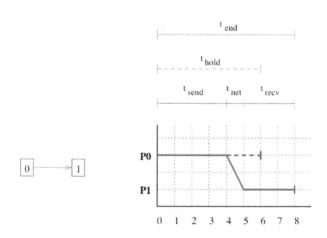

FIGURE 8.4: Broadcasting time costs.

Finally, Park et al. consider two other parameters. t_{end} is the interval between the sender starts sending a message until the receiver finishes receiving. Let t_{hold} be the minimum time interval between two consecutive send operations. The value of t_{hold} is dependent on how blocking send is implemented: it may be greater than t_{end} as t_{hold} may include the overhead of the acknowledgement from the receiver. An example is given in Figure 8.4 and this graph is taken from [Park et al., 1996].

In our approach, we consider that the broadcast operation is based on MPI's broadcast function which is a blocking operation. According to MPI's specification, the blocking operation is considered complete when the sending buffer can be reused.

Optimal broadcast tree for a single cluster

Authors in [Park et al., 1996] have proposed a construction of an optimal broadcast tree based on parameters introduced in the previous paragraph. They propose to proceed as follows.

Let us have a set of k nodes $(P_0, P_1, \ldots, P_{k-1})$, where P_0 is the root node. In the first step P_0 sends to P_j. After t_{hold}, P_0 will be ready to send to an other destination while P_j will be ready to send only after t_{end}. Then the whole set is separated in two subsets rooted at P_0 and P_j where the cardinality of each subset depends on ready time of each corresponding root ($t_0 + t_{hold}$, $t_0 + t_{end}$). And then recursively, each subset is divided in two other subsets in which cardinality depends on ready time of its root. Finally, we define t[i] (for each $1 \leq i \leq k$) as the minimum latency required to broadcast a message among i nodes P_a, $P_{a+1}, \ldots, P_{a+1}$ (for an a, $0 \leq a \leq k-1$) with node P_a as the root node.

Therefore we have:

$$t[i] = \max\{t_{hold} + t[j], t_{end} + t[i-j]\}$$

To ensure the optimality, we must choose the node P_{a+j} such that the broadcast latency is minimal. Thus, according to [Park et al., 1996], we have the following recurrence $t[i]$:

$$t[i] = \begin{cases} 0 & \text{if i=1} \\ \min_{1 \leq i \leq i-1} \{\max(t_{hold} + t[j], t_{end} + t[i-j])\} & \text{if } i > 1 \end{cases}$$

The optimal broadcast tree of a k-node tree is $t[k]$ and this can be computed in $\mathcal{O}(k^2)$ time by using dynamic programming when we explore all possible values of j (the value of the splitter of whole set in two subsets). By establishing a recurrence on splitter's values, Ni et al. could reduce the complexity of the construction of optimal broadcast tree to $\mathcal{O}(k)$.

We shall note that in [Park et al., 1996] all nodes are supposed to be equal. Hence, $t[i]$ denotes the broadcast time in any set of i nodes. The splitter j is chosen only by balancing numbers in the resulting sets. No discussion is done about 'personal' communication performances.

In the introduction part, we have presented some broadcast trees that are currently implemented in libraries. The performance of a broadcast tree is dependent on the number of steps required by the algorithm which is a function of k, the number of participants. The binomial tree requires $\lceil \log_2 k \rceil$ steps. However, in practice, the performance of each broadcast tree is dependent on the software latency of the underlying network. In the reminder, we develop an optimal broadcast tree by considering the two important network parameters t_{hold} and t_{end}. Our approach is similar to the Ni et al. approach [Park et al., 1996]. We extend it in order to take into account a cluster of clusters.

Optimal broadcast tree for a cluster of clusters

As mentioned before for the Grid'5000 testbed, we can suppose that the inter-cluster network is homogeneous and then we can apply the Ni et al. method to build the optimal broadcast tree. In this section we suppose that t_{net} is neglected and t_{end} lies especially on t_{send} and t_{recv}. Subsequently, the Ni et al. approach can be valid in this case. In the real case of Grid'5000, the number of clusters is reduced (9 sites each composed of 200-400 cards), and then the problem can be solved with a dynamic programming approach.

Assume that we have to broadcast a message over the Grid'5000 architecture. The reasoning presented in [Park et al., 1996] stays valid. The broadcast is performed by using proxy on each cluster. Once the message is sent from the root node to the first destination, the whole set of proxies is divided in two subsets as explained before.

To the previous construction we add a local broadcast. Once a proxy receives the message, it has to initiate the broadcast in its local cluster and achieve broadcast in the subset of proxies attributed to it. When it performs the first communication in its subset, the latter is divided in two subsets (as proposed in the work of Ni et al.).

Each proxy is supposed to be equipped with two network cards. The first one is used as an interface with inter-cluster network (RENATER) while the second one is used to initiate the internal broadcast. Hence we can neglect the interference between the inter- and intra-cluster broadcast.

Analytically speaking, when a coordinator receives the message to broadcast then we can estimate the time left to end broadcasting by the maximum of one of these times:

1. the time to broadcast inside its own cluster, noted in the first part of this paper by $T[k]$ (capitalized T);

2. the time to achieve inter-cluster broadcast. The considered coordinator has a subtree of i coordinators rooted at it to broadcast in. According to Ni et al. approach, the time to broadcast in the subtree is recursively divided in two subtrees containing i and $i-j$ coordinators where j is the

appropriate splitter chosen according to t_{hold} and t_{end} of the considered coordinator.

We also note that the notation $t[i]$ is no more valid in our case since broadcast time depends on each coordinator local load. In the remaining we use $T[S]$ where S is a particular set of nodes.

At the beginning, the root, we call it *R1*, initiates the broadcast in its local cluster and sends the message to a given coordinator. The whole tree of coordinators is distributed in two subtrees. Let S be the initial set of coordinator. S is distributed in *S1* and *S2*. *S1* is structured as a tree rooted at *R1* and *S2* is structured as a tree rooted at a given node noted *R2*.

Finally, the broadcast latency can be depicted through the following formula:

$$t[S] = \max\{T[R1], t_{hold} + t[S1], t_{end} + t[S2], \}$$

The broadcast latency is given by the maximum of the local broadcast time, and the broadcast time of *S1* and *S2*. The construction of *S1* and *S2* is done so that we minimize the global broadcast time. And finally

$$t[S] = \begin{cases} T[R1] & \text{if } S = \emptyset \\ \min\{\max(T[R1], t_{hold} + t[S1], t_{end} + T[R2], t_{end} + t[S2])\} & \text{if } S \neq \emptyset \end{cases}$$

If the set of nodes is empty, then the completion time is given by $T[R1]$, the time to achieve local multicast. Otherwise, the set is divided in two. The completion time is given by the maximum of completion time of broadcasting in each subset, and the completion of the local broadcast of both roots. We add parameters t_{hold} and t_{end} to take into account the communication between *R1* and *R2*. *R2* is chosen at each step so that it minimizes the completion time of the remaining steps.

8.4.2 Multi-criteria approach

Introduction

We are now considering the issue of building the optimal broadcast tree from another point of view. Indeed, it is also possible to transform this problem into a bi-criteria problem in the following way.

Let us remember that each cluster contains a number of nodes. We assume that we can build a tree of local broadcast inside each cluster. So, we can determine the time T_i to complete the local broadcast in cluster i. We need now to build the global broadcast tree for all the grid, according to all criteria.

When we reach the first node of a given cluster at time t, it will immediately initiate a local broadcast that takes T_i. We may assume that the (global) broadcast across the grid will not end before $t + T_i$ (i.e., when the most loaded cluster would complete its local broadcast), so since the t time, we have a period of T_i to complete the broadcast in the remaining broadcast sub-tree (i.e., rest of the grid). In others words, we shall say that:

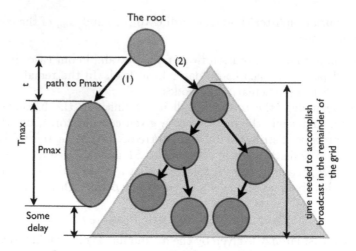

FIGURE 8.5: The most loaded cluster vs. the remaining sub-tree.

- All solutions that would end broadcasting in the remaining sub-tree before $t + T_i$ are equivalent, since the time of completion of the global broadcast will always be the same: $t + T_i$;

- Henceforth, we will always consider in priority the most expensive node in terms of local broadcast: the one that has the greatest value of the T_i term because this node will provide the greatest delay to complete the broadcast in the remaining broadcast sub-tree and therefore the best estimate of the completion time. Let T_{max} be the time needed to broadcast in the cluster P_{max}, the most expensive in terms of local broadcast time. Our completion time is bounded by '*the shortest path to reach to P_{max}* '$+T_{max}$: that is a lower bound and we can never do better.

The term '*the shortest path to reach P_{max}*' means for us the sequence to reach P_{max} as early as possible, either by contacting it directly or by interspersing other nodes to contact it.

So the completion time is $Cmax$ = '*the path to P_{max}*' $+T_{max}+$ '*some delay to finish broadcasting in the remaining of broadcast sub-tree.*' See Figure 8.5 for an example.

We will discuss now the (*path to P_{max}*) and the (*some delay*). The term T_{max} is constant and inevitable because we assumed that the local broadcast is made in an optimal manner. Let us consider two cases:

- (a) If we reduce (the path to P_{max}), i.e., we contact it as soon as possible, we may increase the final delay. Therefore the broadcast tree will look to the remaining broadcasting sub-tree, as shown in Figure 8.6.

- (b) If we delay the contact of the most loaded cluster, i.e., we may reduce the height of the remaining sub-tree; subsequently the considered delay may be reduced (or even avoided). But, on the other hand, we may lengthen the path to reach P_{max}. The broadcast tree will look to the branch that leads to the most loaded node as shown in Figure 8.7.

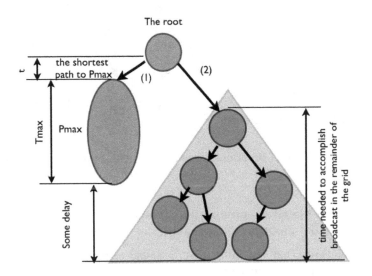

FIGURE 8.6: The broadcast tree looks to the most loaded cluster.

Contribution

According to the discussion given above, the optimal broadcasting time is obtained when the broadcast tree is balanced as much as possible, i.e., we have to reduce the time to reach the most expensive nodes in terms of local broadcast and to reduce the supplementary delay too. These two conditions are complementary and can not be achieved simultaneously.

The optimal solution is obtained when we find the best compromise between these two criteria.

We are therefore faced with two objectives:

1. Reducing the path to P_{max} (case (a) is its optimal solution);

2. Reducing the time between the time when we finish the local broadcast in P_{max} and the end of the global broadcast (case (b) is its optimal solution).

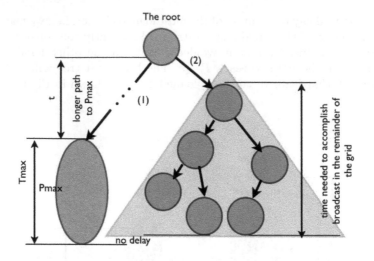

FIGURE 8.7: The broadcast tree looks to the remaining sub-tree.

In a multi-objective optimization [Ehrgott, 2000], [Ehrgott and Gandibleux, 2002] problem we distinguish three kinds of solutions:

  i. the solutions that optimize, first, a single criterion (e.g., case (a) and (b)) and then the other criteria.

 ii. the solutions that are a compromise for all criteria, in a certain way, and that dominate the other solutions. Let S be a solution from this set, then there is no other solution that improves one (or more) criterion without degrading one (or more) other criterion. S is said to be a non-dominated solution or a 'pareto-optimal solution;'

iii. solutions that are dominated. These solutions have no interest since there are other solutions that can improve at least one criteria while respecting its performance for the other criteria.

Remarks

 1. The solutions defined in class i. are also pareto-optimal;

 2. The pareto-optimal class is the most interesting one. It contains all *'optimal'* solutions of a multi-criteria problem. One may list all these solutions and choose the good compromise according to one's will.

Let us come back to our broadcast problem. We have, now, only to explore pareto-optimal solutions between the two solutions (a) and (b). So, we propose the following algorithm for constructing the optimal broadcast tree. The algorithm consists in constructing a balanced tree in an incremental way.

The broadcast tree is initialized by the root (i.e., the origin node of the broadcasting). The set of clusters is sorted in the descending order of T_i. Clusters are re-numbered in this way. The most loaded cluster is numbered $1,\ldots$. The least loaded cluster is numbered N. Clusters are considered in the order from 1 to N. We also need a structure to save branches added to the tree in the order of their insertions.

As the approach is recursive, we consider only the insertion of a single node to the broadcast tree.

The algorithm is as follows:

1. From all the clusters not yet inserted, consider the most loaded cluster. Assume that it is the i cluster. Let T_f be the broadcast time given by the tree before the insertion of the current cluster;

2. Build the path that leads from the broadcast tree to this cluster as early as possible. Note: contact the cluster i as early as possible may be needed for the insertion of other less loaded clusters.

3. Determine the time to achieve the local broadcast in the i cluster. Let T'_f this time. Two cases must be studied:

   - if $T_f \geq T'_f$ then accept the insertion; save inserted branch and proceed to the next cluster in the list.

   - if $T_f < T'_f$ then: (1) note that we reached the inserted cluster i in a time T'_f. (2) We have now to re-balance the tree. Let j be the cluster that gave the time T_f. Remove all branches that lead to the i cluster (the one that gives the time T_f), the branch that lead to the cluster j (the one that gives the T'_f time) and all branches inserted in steps in between. (3) re-insert the cluster j passing the second shortest path and continue insertions. (4) When you come to the first cluster of rank $\geq i$, check that you have not exceeded T'_f; otherwise return to the initial solution.

4. Stop when the cluster's list becomes empty.

   The algorithm tries to satisfy the (a) criterion at the beginning, i.e., minimize the path to the most loaded cluster. But as soon as the criterion (b) is violated (there is a delay, i.e., $T_f < T'_f$) then the algorithm returns on its steps to better balance the broadcast tree while verifying that the more balanced tree does not need more time than the less balanced tree. Thus we have the guarantee that the broadcast time remains as close as possible to the lower bound that we have set at the beginning which is: *the shortest path to reach* $P_{max} + T_{max}$.

8.4.3 Broadcast for clusters

Before the advent of grid, the Network of workstation (NOW) architecture was in vogue in the 90s. NOW is a cost-effective alternative to massively parallel supercomputers based on commercially available off-the-shelf processors to form a cluster. However, a cluster may consist of different types of processors and this heterogeneity within a cluster complicates the design of efficient collective communication protocols.

In [Liu and Sheng, 2000], Pangfeng Liu and Tzu-Hao Sheng show that a simple heuristic called fastest-node-first (FNF) is very effective in reducing broadcast time for heterogeneous cluster systems. Despite the fact that FNF heuristic fails to give the optimal broadcast time for a general heterogeneous network of workstation, they prove that FNF always gives the optimal broadcast time in several special cases of clusters. Based on these special case results, authors show that FNF is an approximation algorithm that guarantees a competitive ratio of 2. From these theoretical results they also derive techniques to speed up the branch-and-bound search for the optimal broadcast schedule in heterogeneous NOW.

The model used in [Liu and Sheng, 2000] depicts a heterogeneous cluster as a collection of processors $P_0, \cdots, P_{n-1}$, each capable of point to point communication with any other processor in the cluster. Since authors are interested in the communication capability of the cluster only, each processor is characterized by its speed of sending messages. A non-negative *transmission time* of a processor represents the time it needs to send a unit of message to any other processor so that the time required to transmit a message is determined by the sender.

The communication model requires that the sender and the receiver processors can not engage in multiple message transmissions simultaneously. Any sender must complete its data transmission to a receiver before sending the next message. Authors assume a one-port model to depict a motherboard with a single network interface for instance. Similarly the model prohibits the simultaneous receiving of multiple messages by any processor.

Let us return to the FNF heuristic. In each iteration the algorithm chooses a sender from the set of processors that have received the broadcast message (denoted by A) and a receiver from the set that have not received the message yet (denoted by B). The algorithm picks the sender s from A so that s will finish this transmission as early as possible, and chooses the receiver r as the processor that has the minimum transmission time in B. Then r is moved from B to A and the algorithm iterates until all the processors are visited once.

Despite its simplicity, FNF does not guarantee optimal broadcast time and an example is given in the aforementioned paper. However, authors demonstrate that FNF is optimal in some special cases of heterogeneous clusters. When there are only two types of processors, authors show that FNF is optimal.

Banikazemi's paper [Mohammad and Dhabaleswar, 1998] is the original paper that has introduced the FNF heuristic and P. Liu uses two theorems in the paper to prove the optimality of FNF in special cases of the heterogeneous system. In [Mohammad and Dhabaleswar, 1998], authors show first that the algorithms such as the Binomial-tree based algorithms which were used for implementing collective operations were not efficient. They propose two new approaches (Speed-Partitioned Ordered Chain (SPOC) and Fastest-Node First (FNF)) to implement collective communication operations with reduced latency.

Authors also investigate methods for deriving optimal trees for broadcast and multicast operations. However, generating such trees is found to be computationally intensive. It is shown that the FNF approach, in spite of its simplicity, can deliver performance within 1% of the performance of the optimal trees. Finally in the paper, these new approaches are compared with the approach used in the MPICH implementation on experimental as well as on simulated testbeds.

The central theorems of [Mohammad and Dhabaleswar, 1998], [Liu and Sheng, 2000] are the following:

THEOREM 8.1
There exists an optimal broadcast tree T in which all processors send messages without delay. That is, for all processor p in T, starting from its ready time, p repeatedly sends a message with a period of its transmission time until the broadcast ends.

THEOREM 8.2
There exists an optimal broadcast tree T in which every processor has a transmission time no less than the transmission time of its parent.

Note that from the definition of optimality in Theorem 8.2, the author considers the optimal broadcast schedule from all possible sources and not from a dedicated source. With Theorem 8.1, we can simply discard those trees that will delay messages, and still find the optimal schedule.

The construction of an optimal broadcast tree is based on a recursive definition of the ready time for a processor to transmit a message. Let $r(s_i)$ be the ready time for processor s_i. In [Liu and Sheng, 2000], authors use the following inductive relation:

$$\begin{cases} r(s_0) = 0 \\ r(s_i) = \min\{t \mid \sum_{j=0}^{i-1}\}NS_S(s_j, t) \geq i\}, 1 \leq i \leq n-1 \end{cases} \quad (8.2)$$

where $NS_S(p, t)$ is the minimum non-negative integer k such that $r(p) + k * t(p) \geq t$, for $t \geq r(p)$. $t(p)$ is the transmission time of a processor p and S the set of processors.

8.4.4   Broadcast and heterogeneous systems

In [Legrand et al., 2005] and also in a technical paper[5] from INRIA, Legrand, Beaumont, Marchal and Robert consider the communication involved by the execution of a complex application deployed on a heterogeneous grid system. Heterogeneous systems are modeled by a graph where resources have different communication and computation speeds. Achieving the best throughput may require that the target system is used in totality.

Authors show that neither spanning trees nor DAGs are as powerful as general graphs. They focus on *series of broadcasts*, meaning that the same source processor sends a series of atomic one-t-all broadcasts. The processing of these broadcasts can be pipelined and the objective function is to optimize the throughput of the steady-state operation, i.e., the average amount of data broadcasted per time unit.

Authors start with a working example and focus on DAGs for describing the network topology. The series of broadcast on a DAG problem is modeled with a linear program and they show that the linear program provides the optimal solution: the value returned by the program is the maximum number of broadcasts that can be initiated per time unit.

Then authors in [Legrand et al., 2005] examine the series of broadcasts on a general system; they mean a system modeled by graphs with cycles. Again they use a linear program and its solution provides a lower bound for the period length needed to broadcast one unit time message. Nevertheless, it is not clear that this bound can be achieved because of the assumption stating that all the messages transiting on a given edge are all subsets of the largest set. Finally, to bypass the difficulty authors use a weighted version of Edmond's branching theorem (see [Schrijver, 2003] vol B, chapter 53) by which they produce the desired number of trees.

8.5   Concluding remarks

In this chapter, we investigated several approaches to achieve a broadcasting operation for a cluster of clusters. The first approach is by approximating the optimal broadcast tree with a new heuristic giving results as good as the best existing heuristic. Known heuristics combine different factors to minimize completion time. Our heuristic does not combine factors but decides at each iteration which elementary factor to satisfy. In the second approach, we defined an exact method to build the broadcast tree. We gave a generalization based on a dynamic-programming method to cluster of clusters. In fact, real

---

[5]See http://www.inria.fr/rrrt/rr-4871.html.

cluster of clusters architecture contains only a few number of clusters. Hence examining all possible broadcasting trees can be achieved in a reasonable time.

We have also pointed out methods based on dynamic programming or based on a multi-criteria approach. In any case, we are studying practical approaches in the sense that we tried to implement them in order to validate them experimentally, as much as possible. We all know that we may encounter a big gap between a nice theoretical result and its implementation that sometimes requires inefficient codes.

In the future, we suggest to the community to fully implement such heuristics inside MPI in order to check whether our predictions are verified on real large scale applications and systems. The challenge is twofold. First of all, to accomplish an experiment is arduous on any large scale system; for instance it is hard to measure the communication time because if you run your experiment on dedicated reserved nodes, the bandwidth is in general shared with others' experiments. In fact, you need to reserve nodes, to deploy your code / kernel and to configure your environment. However, on Grid'5000 we have facilities to accomplish the tasks in a semi-automatic way. Second, we would like to offer a system able to broadcast information according to any message size. So, the challenge is to broadcast on the fly or according to a more 'online' policy.

Last but not least, an orthogonal path is also possible to implement efficient broadcast operations. All routers are now capable of multi-casting which is a synonym for broadcasting. It is quite easy to implement a daemon listening or broadcasting on a dedicated multicast port as it is explained in [Makofske and Almeroth, 2002 , Wittmann, 2000]. The first challenge would be to simulate a `MPI_Bcast` operation using the multicast infrastructure in a VLAN or at a more general level for the network infrastructure. If the first experiments are good, it will be interesting to re-implement MPI collective operation starting with the multicast operation as the basic building block instead of the send/receive operations.

Acknowledgment

Experiments presented in this chapter were carried out using Grid'5000 experimental testbed, an initiative from the French Ministry of Research through the ACI GRID incentive action, INRIA, CNRS and RENATER and other contributing partners (see `https://www.grid5000.fr`). We also thank deeply the Regional Council of Ile-de-France for its support through the SETCI mobility program (see `http://www.iledefrance.fr`).

8.6 References

[Barchet-Steffenel and Mounie, 2006] Barchet-Steffenel, L. and Mounie, G. (2006). Scheduling heuristics for efficient broadcast operations on grid environments. In *Proceedings of the 2006 International Parallel and Distributed Processing Symposium (IPDPS'06)*.

[Barnett et al., 1996] Barnett, M., Payne, D. G., van de Geijn, R. A., and Watts, J. (1996). Broadcasting on meshes with wormhole routing. *Journal Parallel Distributed Computing*, 35(2):111–122.

[Bhat et al., 2003] Bhat, P. B., Raghavendra, C. S., and Prasanna, V. K. (2003). Efficient collective communication in distributed heterogeneous systems. *Journal of Parallel Distributed Computing*, 63(3):251–263.

[Ehrgott, 2000] Ehrgott, M. (2000). *Multicriteria optimization*. Lecture Notes in Economics and Mathematical Systems. Springer-Verlag.

[Ehrgott and Gandibleux, 2002] Ehrgott, M. and Gandibleux, X. (2002). *Multiple criteria optimization: state of the art annotated bibliographic surveys*. Kluwer Academic, Dordrecht.

[Foster and Karonis, 1998] Foster, I. and Karonis, N. (1998). A grid-enabled MPI: message passing in heterogeneous distributed computing systems. In *Proceedings of SuperComputing*. ACM Press.

[Jelena et al., 2005] Jelena, P.-G., Thara, A., George, B., Fagg, G. E., Edgar, G., and Dongarra, J. J. (2005). Performance analysis of MPI collective operations. In *Proceedings of the 19th International Parallel and Distributed Processing Symposium (IPDPS'05)*, page 272.1, Washington, DC, USA. IEEE Computer Society.

[Legrand et al., 2005] Legrand, A., Marchal, L., and Robert, Y. (2005). Optimizing the steady-state throughput of scatter and reduce operations on heterogeneous platforms. *Journal of Parallel Distributed Computing*, 65(12):1497–1514.

[Liu and Sheng, 2000] Liu, P. and Sheng, T.-H. (2000). Broadcast scheduling optimization for heterogeneous cluster systems. In *Proceedings of the 12th Annual ACM Symposium on Parallel Algorithms and Architectures (SPAA'00)*, pages 129–136, New York, NY, USA. ACM Press.

[Makofske and Almeroth, 2002] Makofske, D. B. and Almeroth, K. C. (2002). *Multicast sockets: practical guide for programmers*. Morgan Kaufmann.

[Matsuda et al., 2006] Matsuda, M., Kudoh, T., Kodama, Y., Takano, R., and Ishikawa, Y. (2006). Efficient MPI collective operations for clusters in long-and-fast networks. In *Cluster*. IEEE Computer Society.

[Matsuda et al., 2004] Matsuda, M., Kudoh, T., Tazuka, H., and Ishikawa, Y. (2004). The design and implementation of an asynchronous communication mechanism for the MPI communication model. In *Cluster*, pages 13–22. IEEE Computer Society.

[Mohammad and Dhabaleswar, 1998] Mohammad, B. and Dhabaleswar, K. (1998). Efficient collective communication on heterogeneous networks of workstations. In *Proceedings of the International Conference on Parallel Processing*, pages 460–467. IEEE Society Press.

[Park et al., 1996] Park, J.-Y. L., Choi, H.-A., Nupairoj, N., and Ni, L. M. (1996). Construction of optimal multicast trees based on the parameterized communication model. In *ICPP*, volume 1, pages 180–187, Los Alamitos, CA, USA. IEEE Computer Society.

[Rabenseifner, 1999] Rabenseifner, R. (1999). Automatic profiling of MPI applications with hardware performance counters. In *PVM/MPI*, pages 35–42. Available online at: `citeseer.ist.psu.edu/rabenseifner99automatic.html` (accessed May 1, 2009).

[Ravi, 1994] Ravi, R. (1994). Rapid rumor ramification: approximating the minimum broadcast time. In *Proceedings of the Symposium on Foundations of Computer Science*, pages 202–213, Los Alamitos, CA, USA. IEEE Computer Society.

[Schrijver, 2003] Schrijver, A. (2003). *Combinatorial optimization: polyhedra and efficiency*, volume 24 of *Algorithms and Combinatorics*. Springer-Verlag.

[Steffenel, 2006] Steffenel, L. A. (2006). Modeling network contention effects on all-to-all operations. In *Proceedings of the Conference on Cluster Computing (CLUSTER 2006)*, Barcelona, Spain. IEEE Computer Society. Available online at: `http://hal.archives-ouvertes.fr/hal-00089242` (accessed May 1, 2009).

[Steffenel and Mounié, 2008] Steffenel, L. A. and Mounié, G. (2008). A framework for adaptive collective communications for heterogeneous hierarchical computing systems. *Journal of Computer and System Sciences*, 74(6):1082–1093.

[Tatsuhiro et al., 2004] Tatsuhiro, C., Toshio, E., and Satochi, M. (2004). High-performance MPI. In *Proceedings of the 2004 International Parallel and Distributed Processing Symposium (IPDPS'04)*. Available online at: `http://csdl.computer.org/comp/proceedings/ipdps/2004/2132/02/213220104babs.htm` (accessed May 1, 2009).

[Wittmann, 2000] Wittmann, R. (2000). *Multicast communication: protocols, programming, and applications*. Morgan Kaufmann.

Chapter 9

Load balancing algorithms for dynamic networks

Jacques M. Bahi

LIFC, Université de Franche Comté, IUT Belfort, 2 rue Engel Gros, 90016 Belfort, France

Raphaël Couturier

LIFC, Université de Franche Comté, IUT Belfort, 2 rue Engel Gros, 90016 Belfort, France

Abderrahmane Sider

Computer Science Department, University of Bejaia, Bejaia, Algeria

9.1 Introduction

Distributed computing is the art of using several computing elements (CEs) to solve a given problem. The elements may be parallel shared memory multiprocessor machines, desktop machines, laptops or any electronic devices that can process a set of control and arithmetic instructions.

The aim of distributing a computation is in general to achieve better execution time, but can also be motivated by solving a greater instance of some problems or solving several instances of the same problem each with different data. For other problems, sequential algorithms can even lead to non-reasonable time to be run for certain sizes of the problem so it can be considered untractable. That is why those problems cannot be run with classical microprocessors. Nowadays, achieving parallelism through work distribution happens to be necessary in many scientific fields ranging from simulating fluid

235

molecular dynamics and particle mechanics [Boillat, 1990] to solving large optimization and scientific problems [Bahi et al., 2006b].

Let us focus first on the CEs capabilities. The computation power property is defined by the number of instructions that the CPU elements can execute during a time unit. But the whole computation power of a CE is usually not available for a target application because of operating system processes and user programs running at the same time on it. So the theoretical constant computing power is shared between the target computations and other processes on each CE. As a result, a computation should be able to adapt to the current available computing power of each involved CE. That is why the computation needs to take into account the environment in which it is executed.

In addition, because CEs may have heterogeneous computing power and because it may be difficult for some problems to decompose the work in balanced parts, some CEs may finish their work faster than others. Consequently they remain idle while waiting for the other processors to finish so they can all decide what the next steps of the computation will be. In order to alleviate idling on the one hand and to efficiently use CEs on the other hand, it may be necessary to assign work to CEs proportionally to their available computation power.

The process of deciding when to do it and how to do it is called a load balancing algorithm. Notice that we omit a rather particular case where the aim is only to keep CEs from remaining idle which is termed load sharing. We can see indeed that the former implies the latter but the opposite is not true.

The computing elements usually also communicate data to synchronize their actions and to manage the consistency of the solution being computed. The messages are exchanged between computing elements guided by the computation logic. The messages form a network whose nodes and links are respectively the CEs (or some entities running on them) and communication channels. Besides, channels may be fixed according either to an arbitrary pattern or a logical one. An arbitrary pattern is one that is constrained by the wire or whatever physical platform is used for communication. A logical pattern is rather constrained by the problem/solution entities that actually run in the distributed application. The communication pattern can support or not link failures and message loss. In the first case we say that the communication network is static while in the second we say it is dynamic.

In a static network, the communication network is considered to be static, i.e., no communication channel/link can be broken and no CE can crash or come down and join the computation again. The work on the presented algorithms corresponds to the era of the so called cluster computing which experienced a great spread due to the existence of adequate programming and running libraries like MPI [Gropp et al., 1994] and PVM [Geist et al., 1994]. The computations are often run on high speed local area networks. These two points outline the static network nature of the communication behavior since in general no process/machine can quit and join a computation in PVM or MPI and the network is considered to have predictable behavior during the computation and if it crashes the whole computation is considered crashed. This contrasts with the Internet Computing era in which we are living and which only began some years ago. That explains the move toward load balancing algorithms that support link/node failures.

9.2 A taxonomy for load balancing

Figure 9.1 shows what type of load balancing algorithms can be made according to locality of decisions/migrations and the time they are performed. We can see that these criteria define four types of load balancing. When the decision of how much work is given to each node is made before execution the algorithm is said to be static. If this decision is made at runtime then the algorithm is dynamic.[1] The first column of Figure 9.1 features two static algorithm types one of which is centralized (Global Static Load Balancing, GSLB for short) whereas the other is distributed (Distributed Static Load Balancing, DSLB for short). GSLB algorithms are more common and they have some similarities with scheduling and mapping algorithms. In the second column of Figure 9.1, two other algorithm types are defined: Global dynamic load balancing (GDLB) and Distributed dynamic load balancing (DDLB). These algorithm types are run during execution and allow us to adapt load distribution as the computations proceed.

In the following, we restrict to the dynamic type algorithms which balance load during runtime. The processor which decides which load movements should take place in order to balance the load can be unique or not. If it is unique then the load balancing is said to be centralized or global (GSLB and GDLB classes); otherwise it is distributed or local (DSLB and DDLB classes). LeMair et al. [Willebeek-LeMair and Reeves, 1990] summarized the different techniques that were used in cluster computing oriented environments at the end of the 80s. Another important characteristic of GDLB is that load migration can take place between not necessarily neighboring CEs. Global load balancing algorithms obviously suffer from a bottleneck located at the coordinating processor and from other issues such as network latency because the load is not necessarily exchanged between neighboring nodes. Distributed (DDLB class) algorithms are more robust with respect to the first issue and suffer less from network latency because load information and load itself are

---

[1] Notice that the total amount of load in the system/ditributed computation can be constant or varying with time, i.e., load is consumed/created at runtime. The first case is also termed as "static load" balancing and the second is "dynamic load" balancing but the confusion is usually fixed rapidly as we do here.

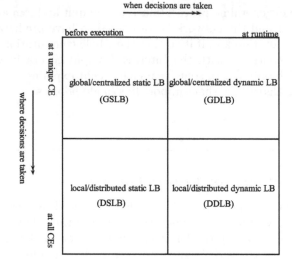

FIGURE 9.1: Load balancing algorithm classes according to locality of decisions/migrations and the time they are performed.

exchanged only locally which means that no routing is needed. We shall restrict to this robust class of algorithms in the following sections.

Another important criterion, within the distributed dynamic algorithms class, is the number of participating nodes that can be less or equal to the total number of entities or processors involved in the computation. In the latter case, the algorithm is said to be synchronous (SGDLB and SDDLB classes of Figure 9.2); otherwise it is asynchronous (AGDLB and ADDLB classes of Figure 9.2). In the synchronous case, all processors stop computations, perform a load balancing step which involves load migration and finally resume computations. The diffusion algorithm, the Dimension Exchange (DE), the Generalized Dimension Exchange (GDE), the second order diffusion all originally run synchronously even if asynchronous implementations have been later investigated. See [Xu et al., 1995] for a comparison between synchronous and asynchronous implementations of Diffusion and GDE.

In the asynchronous scenario, only a subset of the processors does load balancing at a load balancing step. This usually happens when a processor detects that it is underloaded or overloaded[2] with respect to its neighbors or to some fixed threshold values. SID and RID by LeMair et al. [Willebeek-LeMair and Reeves, 1993], DASUD by Cortes et al. [Cortes et al., 1999] are purely asynchronous algorithms, i.e., not adaptations of synchronous versions.

---

[2]The first scenario is termed as receiver initiated load balancing (RID shortly) and the second case as sender initiated (SID shortly).

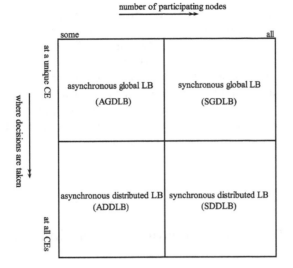

FIGURE 9.2: Dynamic LB algorithm classes according to locality of decisions/migrations and the number of participating nodes.

The Gradient Model by Lin et al. [Lin and Keller, 1987], one of the earlier known algorithms, is a good example of noniterative algorithms of this class.

In SDDLB algorithms, synchronous load balancing iterations are run alternatively with computation phases. During each iteration, load balancing algorithms let some nodes know the available computing power and must tell them which node has to exchange loads with which other nodes and to what amount. These decisions follow some logic that differs from one algorithm to another. The logic may be for example that of the Liquid Model [Henrich, 2004] or some other natural phenomenon. An intensively studied model is that of the physical *diffusion* model that describes how heat spreads through a liquid. What is interesting is that it can be modeled and studied according to the very developed theory of linear algebra because it is iterative[3] in nature. Notice that not every iterative load balancing algorithm can be modeled by a mathematical iterative algorithm hence the important literature on diffusion on static/dynamic homo-/heterogenous networks. A derived form of diffusion called *dimension exchange* has also received a lot of attention due to its closeness to diffusion and to its ability to be implemented on running platforms that do not enable a CE to concurrently communicate with all of its neighbors.[4]

---

[3]In the mathematical acceptation.

[4]In computer systems, this may apply to ones that do not support multiprocessing or multithreading.

This chapter relates to "static load" synchronous distributed dynamic load balancing algorithms namely the SDDLB class of Figure 9.2. But these algorithms can easily be adapted for a "dynamic load" context. We first present their theoretical aspects for static networks in Section 9.3 and then for dynamic networks in Section 9.4. Afterward, Section 9.5 explains practical aspects of integrating a load balancing algorithm into any distributed algorithm. Finally, Section 9.6 shows how load balancing is integrated into a real world scientific application and the obtained gain in static and dynamic network contexts. This chapter ends with a conclusion and some perspectives.

9.3 Distributed load balancing algorithms for static networks

In this section, popular load balancing algorithms for static networks are considered. We focus in particular on two iterative algorithms that have been adapted for dynamic networks namely diffusion and dimension exchange and on their different developments. They participated in improving their performances and in making them able to be run on other static networks than the one they were originally designed for, i.e., the n-dimensional hypercube. But before investigating Diffusion and Dimension Exchange, their variance and their adaptation to dynamic networks, on the one hand, we need first to present the notations used hereafter. On the other hand, because a load balancing algorithm should be assessed like any other distributed algorithm, we should present the measures that permit to judge the effectiveness of any load balancing algorithm. These two points are the focus of the next section.

9.3.1 Network model and performance measures

A distributed-memory parallel system of n processors linked with an interconnection network is modelled by a graph $G = (V, E)$ where vertices V and edges of E represent the processors and links between them, respectively. Note that in a dynamic context, E is constituted only by links that are alive. A link is alive when it can deliver messages in both directions [Aiello et al., 1993]. A link that is not alive is said to be broken. Let E_B^t be the set of broken edges in the graph G at time (iteration) t and $N_i^t = \{j \in V : (i,j) \in E \wedge (i,j) \notin E_B^t\}$ the set of neighbor nodes of processor i at time t. The workload of node i is represented by a nonnegative integer scalar value w_i. At time t the system's load distribution is represented by the vector $W^t = \{w_1^t, w_2^t, w_3^t, \ldots, w_n^t\}$. The target of the load balancing process is to make this load distribution system converge towards the uniform load distribution represented by vector $\overline{W} = \{\overline{w}, \overline{w}, \overline{w}, \ldots, \overline{w}\}$ where $\overline{w}$ is the load that every node should have received if a global knowledge of the overall system's load were known. If the system is built by assembling homogeneous processor powers and link bandwidths then $\overline{w} = \sum_{i=1}^{n} w_i/n$. In the other case, suppose each processors' power is represented by value s_i. Then node i should be allocated a workload which is proportional to its power: that is $\overline{w_i} = \frac{\sum_{i=1}^{n} w_i}{\sum_{i=1}^{n} s_i} s_i$.

Performance metrics

Three fundamental properties are usually considered when assessing the performance of a load balancing algorithm: its termination, its efficiency and its stability [Xu and Lau, 1992]. The notion of termination relates to the ability of the algorithm to lead any initial load distribution to the average load one. This is done mainly by means of formal proofs. The efficiency is a subsidiary result of the termination proof because it shows the execution time a given algorithm spends to reach a load balanced state. If the algorithm is iterative then the rate of convergence is more suitable to assess efficiency. Thirdly, the quality of the obtained global balance is an important criterion because the load is often not so evenly distributed and there still subsists an unbalance between indirectly linked nodes. Similarly to efficiency, if the algorithm is iterative (not direct) than the stability[5] of the algorithm reflects its quality. In practice, this is modeled by some norm of the vector $\bar{W} - W^t$ and is also called the global imbalance. The norm may be the Euclidean max norm noted $l_1 = max_i\{|\ \bar{w} - w_i\ |\}$ or the quadratic one $l_2 = (\sum_i (\bar{w_i} - w_i)^2)^{1/2}$ which will be referred to in this chapter.[6] One important remark should be made here about the cost of load balancing. In static networks, as clarified by Elssser et al. [Elsasser et al., 2002], iterative algorithms of diffusion compute the minimum flux over the network edges. But according to experiments shown in [Sider and Couturier, 2009], this is unfortunately not true in dynamic networks. Indeed, in dynamic networks, the load is eventually balanced by the process but this one does not act according to the heuristic of "minimum flux over network edges" like in static networks. Consequently, the amount of work loads moved over the network edges is an important criterion for a load balancing algorithm especially in dynamic networks since it measures transmission costs and it possibly acts on the efficiency of the algorithm in synchronous implementations.

---

[5]Rate of convergence and stability are the two important properties of any iterative algorithm, in the mathematical acceptation.

[6]It is well known that the max norm, the infinite norm and the Euclidian norm are mathematically equivalent in $\mathbb{R}$.

9.3.2 Diffusion

The diffusion algorithm also known as a First Order Scheme means that the decisions of the load balancing algorithm at time-step t only use load information of the previous time-step $t-1$ and was devised by the pioneering Cybenko in [Cybenko, 1989]. The author supposed a static network communication pattern having a hypercube shape. The hypercube is a particular graph that has many interesting properties. First a hypercube of size n nodes is said to have $D = \ln_2 n$ dimensions. It follows that each of the n nodes has a degree equal to the hypercube dimension D. Neighbor identifiers of a node having identity i are particular in that each one differs only in one bit from that of i. In practice a hypercube of dimension D is built using two hypercubes of dimension $D-1$ and by adding a link from each node of the first hypercube to the node having the same identifier in the other hypercube. Finally, both identifiers are modified by adding one prefix bit and giving it the value 0 for the node of the first hypercube and the value 1 for the node of the other hypercube.

Figure 9.3 shows three examples of hypercubes of dimension one, two and three. The reader should notice how we apply the process that we have described to build the hypercube of dimension two by using two hypercubes of dimension one. We can also check the properties of the hypercubes and in particular the equality of the degree and the diameter.

By the diffusion iterative scheme of Cybenko, on the hypercube static structure, a node i exchanges a portion noted α of its load difference $\mid w_j^t - w_i^t \mid$ with its neighbor j where $\alpha = \frac{1}{\Delta+1}$; Δ being the maximum degree of G. For example, in the hypercube of dimension 3 (cf. Figure 9.3(c)), $\alpha = \frac{1}{3+1} = \frac{1}{4}$. Later, FOS had been optimally tuned for some general static structures called k-ary n-cubes by Xu et al. [Xu and Lau, 1994]. The tuning of FOS is achieved by looking for a diffusion parameter noted α_{opt} that is neither necessarily constant nor equal to $\frac{1}{\Delta+1}$. Moreover, authors gave general formulae to compute it for chain, ring and mesh structures. Boillat's independently devised FOS version [Boillat, 1990] is another approach for choosing the diffusion parameter which becomes dependent in the degrees of two considered processors; i.e., $\alpha_{i,j} = \frac{1}{max(d_i,d_j)+1}$ where d_i and d_j represent degrees of nodes i and j respectively.[7] Anyway, by the FOS diffusion algorithm, the load evolution of processor i at time t is done according to formula 9.1 where $\alpha_{i,j}$ is fixed according to one way from the three above-cited ones.

$$w_i^t = w_i^{t-1} + \sum_{j \in N_i} \alpha_{i,j}(w_j^{t-1} - w_i^{t-1}) \tag{9.1}$$

---

[7]The Boillat diffusion scheme can be seen as a generalization of diffusion to arbitrary networks.

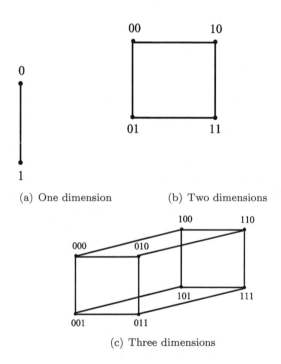

(a) One dimension   (b) Two dimensions

(c) Three dimensions

FIGURE 9.3: The hypercube network of dimension 1, 2 and 3.

By grouping the terms w_i^{t-1} we obtain the following equation

$$w_i^t = (1 - \sum_{j \in N_i} \alpha_{i,j}) w_i^{t-1} + \sum_{j \in N_i} \alpha_{i,j} w_j^{t-1} \tag{9.2}$$

which makes it obvious that the equation 9.1 is linear. The system load evolution between iterations $t-1$ and t can thus be expressed by equation 9.3

$$W^t = MW^{t-1} \tag{9.3}$$

where W^t is the system load vector of size n and M is called the *diffusion matrix* which is defined by its entries m_{ij} such that

$$m_{ij} = \begin{cases} \alpha_{ij} & \text{if } i \neq j, \\ 1 - \sum_{j \in N_i} \alpha_{ij} & \text{if } i = j. \end{cases}$$

The diffusion matrix M is symmetric and doubly stochastic, i.e., $\sum_i m_{ij} = 1$. Cybenko proved that this scheme converges to the uniform load distribution $\overline{W}$. Where the Cybenko and Boillat's schemes offer straight ways to fix α_{ij} which may be the easiest to implement, Xu devised another determination way for α_{ij} and thus for M. In order to find the unique value of α that maximizes the rate of convergence of algorithm 9.3, the eigenvalues of the diffusion matrix should be computed. The second largest of them noted γ plays the most important role in the rate of convergence since this is equal to $-\log \gamma$. To make the convergence rate the fastest, γ should be minimized. This method is not direct and needs to be conducted for every topology. Nevertheless, several studies on some classical topologies have been carried, thus:

$$\begin{array}{ll} \alpha = \frac{1}{2} & \text{for a chain} \\ \alpha = \frac{1}{2n} & \text{for a mesh } k_1 \times k_2 \times \ldots k_n \\ \alpha = \frac{1}{2n+1-\cos(\frac{2\pi}{k})} & \text{for a torus } k_1 \times k_2 \times \ldots k_n \text{ where } k = \max k_i \\ \alpha = \frac{1}{n+1} & \text{for an hypercube of dimension } n \end{array} \tag{9.4}$$

Let us illustrate with more details the functioning of FOS. According to the methods we have just presented the diffusion matrix M may be determined in function of the degree of each node, in function of the degree of the graph or for an optimal convergence rate. For the hypercube of dimension 3 of Figure 9.3(c), a regular graph (same degree for all nodes), we can see that the the three methods coincide for the choice of α and thus the same diffusion matrices are found. Consider now the arbitrary network of Figure 9.4.

Table 9.1 shows the obtained diffusion matrices with Boillat's method (a) and with Cybenko's method (b).

For Xu's method, positive entries of both matrices should be replaced with α except diagonal entries for which we have $1 - x\alpha$ where x is the number of nonzero entries of the considered row. The matrix M is usually written

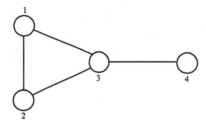

FIGURE 9.4: An arbitrary network of size 4.

(a) With the Boillat's method.

$$\begin{bmatrix} 5/12 & 1/3 & 1/4 & 0 \\ 1/3 & 5/12 & 1/4 & 0 \\ 1/4 & 1/4 & 1/4 & 1/4 \\ 0 & 0 & 1/4 & 3/4 \end{bmatrix}$$

(b) With Cybenko's method.

$$\begin{bmatrix} 1/2 & 1/4 & 1/4 & 0 \\ 1/4 & 1/2 & 1/4 & 0 \\ 1/4 & 1/4 & 1/4 & 1/4 \\ 0 & 0 & 1/4 & 3/4 \end{bmatrix}$$

(c) With Xu's method.

$$\begin{bmatrix} 1-2\alpha & \alpha & \alpha & 0 \\ \alpha & 1-2\alpha & \alpha & 0 \\ \alpha & \alpha & 1-3\alpha & \alpha \\ 0 & 0 & \alpha & 1-\alpha \end{bmatrix}$$

Table 9.1: Diffusion matrices for the network of Figure 9.4.

in function of the Laplacian L of the considered graph in order to take into account the graph shape, on the one hand, and to facilitate the task of computing α_{opt}, on the other hand. Indeed, $M = I - \alpha L$ where $L = AA^T$ is the Laplacian, A is the incidence matrix of the graph, A^T its transpose and I the identity matrix of size n. The convergence rate of diffusion is a function of γ, the second largest eigenvalue in absolute values of the diffusion matrix eigenvalues: $\gamma = \max|\mu_2|, |\mu_n|$ ($1 = \mu_1 > \mu_2 > \ldots > \mu_n > -1$). The rate of convergence is precisely equal to $-\log\gamma$ with $0 < \gamma < 1$ and consequently $\log\gamma < 0$ and $-\log\gamma > 0$. The convergence rate is maximum when γ is minimum. Because of the relation between M and L, we have $\mu_n = 1 - \alpha\lambda_2$ and $\mu_2 = 1 - \alpha\lambda_n$ where λ_2 and λ_n are respectively the second smallest and second largest of the Laplacian eigenvalues ($0 = \lambda_1 < \lambda_2 < \ldots < \lambda_n$). γ is minimum when $\alpha_{opt} = \frac{2}{\lambda_2 + \lambda_n}$ [Xu and Lau, 1996].

So far we have not dealt with communication links bandwidths that may be heterogenous, i.e., not equal, yet. The diffusion algorithm has been adapted to this context (with homogenous computing powers) by Diekmann in [Diekmann et al., 1999]. Finally, diffusion in heterogenous CEs and communication links was studied by Elsasser et al. [Elsasser et al., 2002] and independently by Rotaru et al. [Rotaru and Nageli, 2004] who put the latest developments of diffusion on static networks.

9.3.3   Dimension exchange

Another approach for direct neighbor load balancing is to let a processor exchange some load with only one of its neighbors at each load balancing step. Cybenko in [Cybenko, 1989] presented the DE (*Dimension Exchange*) algorithm on the structure of a hypercube interconnecting network. Under this scheme, a processor with id i will exchange some load with some other node j (and vice versa) on a dimension of the hypercube noted d and computed according to formula $d = (t \bmod D) + 1$ where t and D are respectively the current iteration number of the load balancing algorithm and the number of dimensions of the hypercube; hence the name of the algorithm. Indeed each dimension of the hypercube is a set of edges (whose extremities are pairs of CEs) so that each one can be noted E_d. Formula 9.5 shows how the load of processor i evolves between iteration $t - 1$ and t.

$$w_i^t = w_i^{t-1} + \frac{1}{2}(w_j^{t-1} - w_i^{t-1}) \text{ if (i,j)} \in E_d \text{ and } d = (t \bmod D) + 1 \qquad (9.5)$$

By doing the same work of grouping w_i^{t-1} terms we also finish with the linear equation

$$W^t = M_d W^{t-1} \qquad (9.6)$$

where W^t is the system load vector of size n and M_d is called the *diffusion matrix* of dimension d which is defined by its entries m_{ij} such that

$$m_{ij} = \begin{cases} \frac{1}{2} & \text{if } i \neq j \text{ and } (i,j) \in E_d, \\ \frac{1}{2} & \text{if } i = j \text{ and } (i,j) \in E_d, \\ 0 & \text{elsewhere.} \end{cases}$$

Let us illustrate the functioning of DE with an hypercube of dimension 3. Figure 9.5 shows what are the links on which load exchange takes place at iterations 0, 1 and 2. For example, at iteration $t = 0$ CE with id O (000 in the figure) computes $d = 0 \bmod 3 + 1 = 1$ and CE with id 1 (001) finds equally $d = 1$ which means that their ids differ only in bit number one from the left. So node 0 and 1 can compute the *id* of their partner in the load exchange by simply inverting the dth bit of their *ids*. The same reasoning is applied by other nodes at iteration 0 to find the *id* of their pair (so at iteration 0 nodes along a vertical link can exchange some of their loads). At iteration $t = 1$ however, node O finds $d = 2$ which means that its pair *id* is different by the second bit from the left of its own *id* thus calculating $010 - 2$ so the edge (0,2) will serve for load exchange. By this same reasoning we will have load exchange on edges (1,3), (4,6) and (5,7) (so at iteration 1, nodes along an in-depth link can exchange some of their loads). At iteration 2, nodes along an horizontal link can exchange some of their loads. We can notice that after three iterations load is exchanged on all dimensions, as illustrated in Figure 9.5. Cybenko [Cybenko, 1989] stated that the DE algorithm will converge after exactly $t = D$ iterations for an hypercube of dimension D whatever the initial load distribution is.

Later on, Hosseini et al. [Hosseini et al., 1990] generalized DE for arbitrary interconnection networks by simulating the D dimensions with the maximum number of K colors necessary for coloring the edges of the graph representing the network topology. It is well known that their minimum number is bounded with the maximum degree of the graph Δ: $\Delta \leq K \leq (\Delta + 1)$ [Fiorini and Wilson, 1978]. Load exchange takes place between two nodes i and j iff i and j are the extremities of some edge $e \in E$ having color $k = t \bmod K + 1$. However, with DE and its generalized Hosseini version, the portion of load difference that is effectively exchanged is $\lambda = \frac{1}{2}$ and for this reason this scheme is called ADE (*Averaging Dimension Exchange*). If we note by E_k the edge set of color k, then for a CE i, the load evolves according to formula

$$w_i^t = w_i^{t-1} + \frac{1}{2}(w_j^{t-1} - w_i^{t-1}) \text{ if (i,j)} \in E_k \text{ and } k = (t \bmod K) + 1. \quad (9.7)$$

Again the system load evolves according to the linear equation

$$W^t = M_k W^{t-1} \quad (9.8)$$

where W^t is the system load vector of size n and M_k is called the *diffusion*

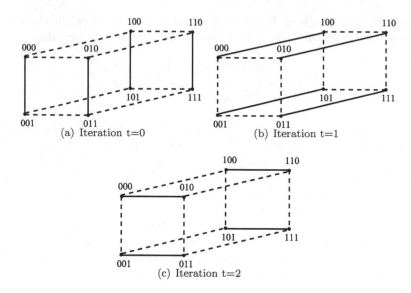

FIGURE 9.5: DE running on an hypercube of dimension 3. At each iteration, CEs use a new dimension for load balancing.

matrix of color k which is defined by its entries m_{ij} such that:

$$
m_{ij} = \begin{cases} \frac{1}{2} & \text{if } i \neq j \text{ and } (i,j) \in E_k, \\ \frac{1}{2} & \text{if } i = j \text{ and } (i,j) \in E_k, \\ 1 & \text{if } i = j \text{ and } (i,j) \notin E_k, \\ 0 & \text{elswhere.} \end{cases}
$$

Notice that with ADE some diagonal entries of M_k are set to 1 simply because a node has not all the K colors on its incident edges. Notice also that unlike DE, that converges after exactly D iterations on the hypercube, ADE also converges on arbitrary networks but usually after more than K iterations.

To illustrate ADE on arbitrary networks, we come back to Figure 9.4. Before running ADE, we must color the graph. In this case we need at most 3 colors. Figure 9.6 shows one possible edge-coloring of the arbitrary network of Figure 9.4. The colors are 1, 2 and 3 so we have three edge sets $E_1 = \{(1,2),(3,4)\}$, $E_2 = \{(1,3)\}$ and $E_3 = \{(2,3)\}$ and three diffusion matrices M_1, M_2 and M_3 that are shown in Table 9.2.

9.3.4 GDE

With GDE (*Generalized Dimension Exchange*) [Xu and Lau, 1992] or ODE (Optimally tuned DE), Xu et al. proved that setting $\lambda = \frac{1}{2}$ would lead to the maximum convergence rate only on the hypercube. As for ADE for arbitrary

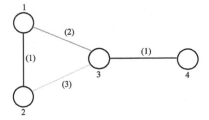

FIGURE 9.6: An edge-coloring of the arbitrary network of Figure 9.4.

$$M_1 = \begin{bmatrix} .5 & .5 & 0 & 0 \\ .5 & .5 & 0 & 0 \\ 0 & 0 & .5 & .5 \\ 0 & 0 & .5 & .5 \end{bmatrix} \quad M_2 = \begin{bmatrix} .5 & 0 & .5 & 0 \\ 0 & 1 & 0 & 0 \\ .5 & 0 & .5 & 0 \\ 0 & 0 & 0 & 1 \end{bmatrix} \quad M_3 = \begin{bmatrix} 1 & 0 & 0 & 0 \\ 0 & .5 & .5 & 0 \\ 0 & .5 & .5 & 0 \\ 0 & 0 & 0 & 1 \end{bmatrix}$$

Table 9.2: Diffusion matrices of the colored graph 9.6.

networks, GDE needs an edge-coloring. They go on and give an alternative way to fix the optimal value λ_{opt} for the k-ary n-cube structures like the chain, the ring, the mesh and the torus. Formula 9.9 expresses how node i load evolves at each timestep t

$$w_i^t = w_i^{t-1} + \lambda_{opt}(w_j^{t-1} - w_i^{t-1}) \tag{9.9}$$

where

$$\lambda_{opt} = \tfrac{1}{2} \qquad \text{for an hypercube,}$$

$$\lambda_{opt} = \frac{2 - \sqrt{2(1 - \cos(\frac{2\pi}{n}))}}{1 + \cos(\frac{2\pi}{n})} \begin{cases} \text{for a } 2n_1 \times 2n_2 \text{ torus where } n = \max n_1, n_2, \\ \text{for a ring of size } 2n, \\ \text{for a mesh } n_1 \times n_2 \text{ where } n = \max n_1, n_2, \\ \text{for a chain of size } n. \end{cases}$$

$$\tag{9.10}$$

The difference with ADE is that entries of any M_k that correspond to an edge of color k now will have a unique λ_{opt} as value. The diagonal entry of each row is then computed so that we express that the load is conserved by the load balancing algorithm. The application of k successive load balancing steps results in the system load evolving to

$$W^{(t+k)} = M_{k-1} \times M_{k-2} \times \cdots \times M_0 W^{(t)} = M(\lambda)W^{(t)},$$

where $M(\lambda)$ is called the GDE matrix. $M(\lambda)$ corresponds to the diffusion matrix M of the diffusion algorithm. The load distribution converges at a rate of $-\log\gamma$, hence the optimal value of λ, λ_{opt} for a graph with Laplacian L is $\lambda_{opt} = \frac{2}{\lambda_2 + \lambda_n}$ where λ_2 and λ_n are respectively the second smallest and second largest eigenvalues of L (see [Xu and Lau, 1996] for details).

With DE for the hypercube a given processor makes load balance operations successively with all of its neighbors. The order is fixed according to the dimension. But with ADE or GDE for arbitrary networks, the order is fixed according to color index k.

9.3.5   Second order algorithms

A second order diffusion scheme (SOS hereafter) is simply a diffusion load balancing algorithm that uses load information of timestep $t - 1$ and that of timestep $t - 2$ to make a decision on the amount of load to exchange at timestep t. Diekmann et al. [Diekmann et al., 1999] introduced the second order diffusion. The scheme relies on a ponderation parameter β. At timestep t, the load of node i exchanging load with neighbor j evolves as described in equation 9.11.

$$w_i^1 = w_i^0 + \alpha_{ij}(w_j^0 - w_i^0) \text{ if } t = 1$$
$$w_i^t = (\beta - 1)(w_j^{t-2} - w_i^{t-2}) + \beta\alpha_{ij}(w_j^{t-1} - w_i^{t-1}) \text{ if } t \geq 2. \quad (9.11)$$

The SOS diffusion algorithm converges if and only if $\beta \in]0, 2[$ by a result from Varga [Varga, 1962]. Additionally, care should be taken in setting the β value because the load balancing iterative algorithm has the additional constraint that the variables w_i^t must remain nonnegative (see [Vernier, 2004] page 64 for details).

Another second order diffusion scheme called the Chebyshev scheme has been introduced by Gosh et al. in [Ghosh et al., 1996]. It has the property that β depends in timestep t and is computed according to equation 9.12 where μ_2 is the second largest eigenvalue of the diffusion matrix M. As for SOS, care should be taken when setting β^t (ibid.).

$$\beta^1 = 1 \text{ if } t = 1$$
$$\beta^2 = \frac{2}{2 - \mu_2^2} \text{ if } t = 2$$
$$\beta^t = \frac{4}{4 - \mu_2^2} \text{ if } t \geq 3 \quad (9.12)$$

9.4   Distributed load balancing algorithms for dynamic networks

As we previously mentioned, a dynamic network is a model for a communication network where some links may be down and be recovered or some messages lost but the nodes are not allowed to crash or to come down and recover. FOS and GDE have received a lot of attention for their simplicity and solid theoretical bases in static networks and so just naturally for dynamic ones.

9.4.1 Adaption to dynamic networks

The adaptation to dynamic networks of diffusion and dimension exchange is based on the hypothesis that the nodes cannot crash and recover but the links can. So, this is the model for a grid computing infrastructure where some communication channels can become unavailable for physical or logical crash. A logical link failure is for instance a network contention that leads to not delivering application messages carried out by the TCP (Transfer Control Protocol) transport protocol.

Bahi et al. [Bahi et al., 2003b] adapted DE for the hypercube with broken edges. The adaptation is done by giving the necessary and sufficient conditions that make an algorithm "good" for a dynamic network. For instance with DE, the algorithm proceeds by dimensions such as in static networks but if an edge is broken then it is simply not used. Later on, the general scheme for adapting GDE for dynamic networks of any initial shape was sketched by Bahi et al. in [Bahi et al., 2005] and named GAE (*Generalized Adaptive Exchange*). We will come back to it in the next section more in depth. Diffusion algorithms have also been studied in this new network model. We will just cite important literature about diffusion on dynamic networks by Bahi et al. [Bahi et al., 2005] and Elsasser et al. [Elsasser et al., 2004]. Finally, Relaxed FOS algorithm (called RFOS) [Bahi et al., 2003a] for dynamic networks has proven to be faster than FOS. DE, GAE, FOS and RFOS on dynamic networks rely on a simple yet very realistic hypothesis that no part of the network remains disconnected during the application of the load balancing algorithm.

The analytical proof for load balancing termination on dynamic networks relies on the definition of the communication graph for load balancing at time t noted G^t. This graph contains all the edges that are used for load balancing at time t. The superposed communication graph for load balancing between times t and $t + n$, noted $G^{t,t+n}$, is then defined by the superposition of the communication graph from time t to time $t + n$. The necessary and sufficient condition for a load balancing algorithm to reach the load balance state on dynamic networks is as follows: If for every t it exists an instant $t + n$ such that the superposed communication graph $G^{t,t+n}$ is connected, then the load balancing algorithm converges (please refer to [Bahi et al., 2003a] for the proof; Vernier [Vernier, 2004] also gives a detailed proof in French).

9.4.2 Generalized adaptive exchange (GAE)

The GAE (*Generalized Adaptive Exchange*) [Bahi et al., 2005] is the adaptation of GDE for dynamic networks in which broken edges are considered. There are many differences between GDE and GAE. First, GDE is based on an edge-coloring while GAE is not. For a given node, at time t, a node does not balance its load on an edge with color k but it balances its load with any neighbor provided that this neighbor is not balancing its load with another node at the same time. This modification is essential in the context of a dynamic network. As an example, suppose that using GDE a processor i must balance its load with its neighbor j on the edge of color k. If the edge is broken then i will not be able to balance its load with j. But with GAE, node i can balance its load with another node, say h, provided that the link (i, h) is not broken and h is not balancing its load with one of its neighbors. With GAE, load evolution of CE i evolves according to equation 9.13.

$$
\begin{aligned}
w_i^t &= w_i^{t-1} + \lambda(w_j^{t-1} - w_i^{t-1}) \text{ if } i \text{ communicates with a} \\
&\quad \text{neighbor } j \text{ at time } t \\
&= w_i^t \text{ if } i \text{ does not communicate with} \qquad\qquad (9.13) \\
&\quad \text{any neighbor at time } t
\end{aligned}
$$

Authors showed that this can be done according to different policies, all of which are only a particular case of FOS on dynamic networks and hence the convergence of GAE is guaranteed under the same assumptions. The outcome of these policies is a set of neighboring nodes pairs. The next few sections explain briefly the different policies and more extensively the one that is named most to least loaded.

9.4.2.1   Basic strategies

GAE can be performed according to three different policies named arbitrary, random and most to least loaded (M2LL). The outcome of each of these policies is a set of pairs (two neighbor nodes) that should exchange load. The arbitrary policy consists in letting each CE choose a pair on the basis of its highest/lowest identifier or by a round robin order (the edge-coloring is simply a particular implementation of it). By the random policy, however, the process of choosing a pair is probabilistic because a node randomly selects one of its neighbors. This policy is similar to Random Polling which has been broadly investigated by Sanders [Sanders, 1994] but only in static networks.

Finally, the most to least loaded policy (M2LL) consists in choosing as pairs the two nodes of each neighborhood domain that have the greatest imbalance. By tackling these imbalances precisely, the global imbalance is reduced in its most important terms which in turn leads very fast to the balance state.

9.4.2.2   Most to least loaded policy (M2LL)

The outcome of M2LL is a finite set of pairs such that each one represents two processors which have the greatest unbalance of their neighborhood. By making these two nodes exchange some load, the objective is to tackle all such unbalances and thus to reach a balanced state as quickly as a pair can be formed. The GAE algorithm together with the M2LL policy can be seen as a per-iteration implementation of GDE. In fact, M2LL creates, when it is invoked (generally before applying GAE), a virtual ad hoc coloring in which nodes of each dimension are chosen in such a way, that the load they will exchange will lead to a fast convergence to the uniform load distribution. Fixed pairs are different from one iteration to another according to load differences between nodes and available alive links.

M2LL distributed algorithm (illustrated in algorithm 9.4.1) uses many concepts and proceeds by stages. A short explanation will be given here; for a full description please see [Bahi et al., 2006a]. The concepts are the decision of a processor, the load order, the interest for load balancing, the best interest, the most interesting processor, the degree of freedom of an interesting processor and the preference of a processor. Each M2LL iteration has exactly four stages. The first stage is simply load information exchange with available neighbors on living links. The second stage is named interest exchange during which the best interest of each processor is exchanged within its neighborhood N_i^t at timestep t. After that stage, each processor determines the set of its best interesting processors B_i^t. The best interesting processor is searched by the preference function which acts on B_i^t or on the set of less loaded interesting processors $L_i^t \subseteq B_i^t$ if node i has the so-called "problem of centered load." This search and exchange constitute the third stage. Finally, the best interesting processor identifier is exchanged between nodes of each B_i, the pair processor is eventually found by each participating node and the new status of each processor is exchanged within N_i^t. The notion of decision in conjunction with the existence of a pair is equivalent to the relation "node i communicates with node j" that is used in the definition of the GAE algorithm [Bahi et al., 2005].

In the first stage (lines 4-5), each processor exchanges load information *locally* on alive links then keeps iterating (lines 10-27) within M2LL until it finds a pair or knows that all its neighbors took their decision. During one M2LL sub-iteration, a processor exchanges two kinds of messages: interest exchange and decision messages. On line 13, the outcome of the preference function is sent to adjacent nodes that are "not decided yet." Finding the most interesting processor allows each node to eventually find a pair. The necessary condition is stated in line 16. In the last stage (lines 21-22), each processor indicates to its neighbors, participating in the current M2LL sub-iteration whether it has found a pair by a decision message. If so, its neighbors that have not succeeded in taking their decision after the current sub-iteration, remove it from their respective load orders.

Experimental results showed clearly that although GAE with M2LL policy suffers from migration communication cost that should be reduced, it is still faster than RFOS in dynamic networks and achieves a good quality of balance comparable with that of DE or GDE on static networks.

Algorithm 9.4.1 Generalized Adaptative Exchange (GAE) with the M2LL policy

---

1: $decided_i^t(i) = false$;
2: $Pair_i^t = UNKNOWN$; {GAE start}
3: **for all** $j \in N_i^t$ **do**
4:     send(w_i^t, j);
5:     receive(w_j^t);
6: **end for**{exchange load information with all living links}
7: bool $localBalance_i^t = \forall j \in N_i^t : |w_j^t - w_i^t| \leq 1$;
8: **if** ($localBalance_i^t = false$) **then**
9:     {M2LL start}
10:     **while** $\neg decided_i^t(i)$ **do**
11:         Find the processor $MostInteresting_i^t$;{with the preference function}
12:         **for all** $j \in N_i^t$ such that $\neg decided_i^t(j)$ **do**
13:             send($MostInteresting_i^t, j$);
14:             receive($MostInteresting_j^t$);
15:         **end for**
16:         $Pair_i^t = j \iff \exists j \in N_i^t : \neg decided_i^t(j) \wedge MostInteresting_i^t = j \wedge MostInteresting_j^t = i$ {Find $pair_i^t$}
17:         **if** ($Pair_i^t \neq UNKNOWN$) **then**
18:             $decided_i^t(i) = true$;
19:         **end if**
20:         **for all** $j \in N_i^t$ such that $\neg decided_i^t(j)$ **do**
21:             send($decided_i^t(i), j$);
22:             $decided_i^t(j)$=receive($decided_j^t(j)$);
23:         **end for**
24:         **if** ($Pair_i^t = UNKNOWN$) **then**
25:             $decided_i^t(i) = \exists j \in N_i^t : \neg decided_i^t(j)$;
26:         **end if**
27:     **end while**
28:     {M2LL end}
29:     **if** ($decided_i^t(i) = true$) **then**
30:         **if** ($Pair_i^t = j \neq UNKNOWN$) **then**
31:             $w_i^{t+1} = w_i^t + \lambda(w_j^t - w_i^t)$;
32:         **else**
33:             $w_i^{t+1} = w_i^t$;
34:         **end if**
35:     **end if**{GAE end}
36:     migrate-load();
37: **end if**

9.4.3 Illustrating the generalized adaptive exchange most to least loaded policy on a dynamic network

In order to show more clearly how the GAE M2LL algorithm balances the load in a dynamic network, let us proceed with an example. Figure 9.7 shows in 9.7(a) the initial static network, nodes *ids* are noted outside and their load is inside. We will follow the synchronous execution of GAE using a standard exchange value $\lambda = \frac{1}{2}$ even if an optimal value will be faster as was reported in [Sider and Couturier, 2009].

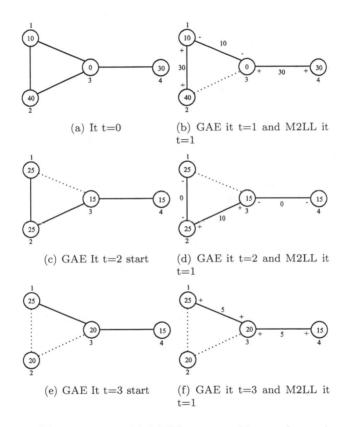

(a) It t=0

(b) GAE it t=1 and M2LL it t=1

(c) GAE It t=2 start

(d) GAE it t=2 and M2LL it t=1

(e) GAE It t=3 start

(f) GAE it t=3 and M2LL it t=1

FIGURE 9.7: GAE running with M2LL on an arbitrary dynamic network of 4 nodes.

At the beginning of every GAE iteration, the broken edges are supposed to be known. For example in 9.7(b) the edge $(2,3)$ is broken. On every edge, we can see the interest of each node extremity for the other node extremity. The interest is a measure of the difference in load between two processors i, j

and is equal to $|w_j - w_i|$. For instance in Figure 9.7(b), the interest of node 3 for node 4 is 30 load units. Likewise $Interest_1(3) = 10$, $Interest_3(1) = 10$ and $Interest_1(2) = Interest_2(1) = 30$. The inbalance of a neighborhood is captured by the best interest for load balancing which is simply the maximum interest for a node. For example, for node 1 in Figure 9.7(b), $BestInterest_1 = \max\{10, 30\} = 30$, for node 3 $BestInterest_3 = \max\{10, 30\} = 30$, likewise $BestInterest_2 = \max\{30\} = 30$ and $BestInterest_2 = \max\{30\} = 40$.

At this stage every node knows the greatest imbalance in its neighborhood and its imbalance with each of its neighbors so we can say that it is interested in definitive by its neighbors that have the greatest imbalance; they form the set of its interesting processors. For example, node 1 is interested by node 2 (the + sign in the figure) but not by node 3 (the - sign). Therefore $B_1 = 2$ and $B_2 = 1$. Similarly, $B_3 = 4$ and $B_4 = 3$. Only some steps are now necessary to each node to find its pair. The pair of each node is searched for in the set of its interesting processors. Before deciding on a pair, a single step called the finding of the most interesting processor is performed. By this operation, each node i selects one processor from B_i as a candidate pair. If B_i is a singleton then things are rather simple; the only interesting processor is the most interesting processor for i[8]. This happens to be the case for our nodes in this example so $MostInteresting_1 = 2$, $MostInteresting_2 = 1$, $MostInteresting_3 = 4$ and $MostInteresting_4 = 3$. After this selection is made, its result is exchanged between all nodes of the union of B_is. Now, M2LL is about to reach its first iteration. Node 1 pair is 2 and $(3, 4)$ is the other pair. M2LL finishes and GAE can proceed. Load exchange results in a new load distribution depicted in Figure 9.7(c). Notice that the global inbalance that was $\frac{1}{4} \cdot \sqrt{(10 - 20)^2 + (40 - 20)^2 + (0 - 20)^2 + (30 - 20)^2} \approx 7.9$ is reduced to 2.5 at the beginning of the second GAE iteration.

Indeed Figure 9.7(c) also shows that the network topology has changed at the start of the GAE iteration 2. The link $(2, 3)$ is now alive while $(1, 3)$ is broken. By following the same steps as we do for GAE iteration 1, nodes search anew for pairs in M2LL iterations and again only one is needed in order for all nodes to find a pair or to know that all their neighbors have taken their decision (to find a pair or not to find a pair because all neighbors have pairs). Figure 9.7(f) shows a particular load situation. Processor 3 has a problem of centered load. Therefore he has to choose its most interesting processor in L_i which is the set of its underloaded interesting processors. At the end of M2LL iteration 1, the pair $(3, 4)$ is formed and processor 3 informs CE 2 that it has found a pair. In M2LL iteration 2, processor 2 finds that all its neighbors are either on broken edges or have already (in previous M2LL iterations) found a pair. It considers all of them as they have taken their decision. In its turn, it can take its own decision : it will not have a pair for this GAE iteration. So

---

[8]If B_i is not a singleton then the preference function is used to select one by a sophisticated mechanism.

the distributed M2LL algorithm terminates after 2 iterations in this case. We step over particular details of M2LL distributed algorithm like the use of the degree of freedom of an interesting processor. The interested reader is invited to refer to [Bahi et al., 2006a] for more details.

9.5   Implementation

The task of implementing a load balancing scheme and integrating it into a distributed computation is not a trivial one. Fortunately, numerous libraries and tools have been developped to simplify this task.

Figure 9.8 shows a generic model for distributed computations integrating a load balancing algorithm and running on a Grid platform. This picture is important in order to be aware of all the practical possibilities under hand. Third party libraries for monitoring system performance such as SIGAR or simple system unix API "pstat" (SYSTAT for linux) functions, in addition to the grid middleware interface (DIET for example implements Chandra and Toueg and Aguilera failure detector), can be of great help for efficient implementations.

In this section, we will focus on precise aspects of implementing load balancing algorithms and their integration. After some considerations about synchronous and asynchronous implementations, we will discuss in particular diffusion and generalized dimension exchange implementations on static networks and corresponding algorithms for dynamic networks, RFOS and GAE.

Nevertheless, before implementing a load balancing algorithm in an application, it is mandatory to measure the impact that the load migration will have over data. In fact, it may be the case that load balancing alters the topology of the communication. In the following we give a simple example. Consider the following computation with a 3x3 block matrix where every node sends/receives data to/from the nodes upward, downward, to the left and to the right. So node 5 needs to communicate with nodes 2, 8, 4 and 6 (as illustrated in Figure 9.9(a)). If node 5 sends a part of its load to processor 6, then node 6 needs to communicate with nodes 2 and 8 (as shown in Figure 9.9(b)).

9.5.1   On synchronous and asynchronous approaches

When dealing with synchronous load balancing we should decide on the load balancing frequency which defines how often the load balancing algorithm should be run during the computation. This frequency has to be carefully chosen because of the LB algorithm overhead and to other considerations. Indeed, on the one hand, the overhead should be minimized and, on the other

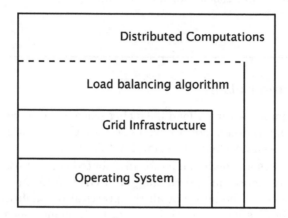

FIGURE 9.8: A generic architecture for distributed computations with load balancing on Grids.

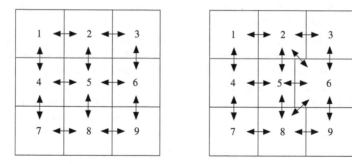

◄──► Communication between 2 processors

(a) Scheme of communications before (b) Scheme of communications after
load balancing load balancing

FIGURE 9.9: Illustration of a modification of the topology of communications due to a load migration.

hand, the load balance state should be reached so that we earn the fruit of integrating the LB algorithm which resumes definitely at minimizing the total execution time. In addition, the available power computation may vary due to other computations/background process in the system. So load balancing should take into account this variability as frequently as possible which results in an argument for a high frequency. In conclusion an aggressive approach may affect negatively the computation while a lazy policy will not balance the computation and will waste a potential improvement.

In the asynchronous approach, low and high thresholds should be defined. They will help a CE to detect whether it is underloaded or overloaded. Similarly to the exchange parameter value of GDE in arbitrary networks synchronous implementations, setting the thresholds' values remains a significant aspect that will decide *in fine* on the gain in performance of asynchronous implementations. Once again, the test and run approach is the only possible way to carefully set the thresholds.

Finally, both in synchronous and asynchronous approaches, a migration function/component is to be implemented. This forms an interface between the LB function/componenent and the computation function/component. Its role is to actually send or receive data and to update local information in relation to the computation so that the computations can safely go on.

9.5.2   How to define the load for some applications

A precise measure of the available power of each CE participating in the computation is a good starting point for a successful implementation. Some existing libraries or an ad hoc measure (such as the CPU usage or the CPU queue length) available in many operating systems catalogs and accessible to any process through system API calls can do the job. Then the load should be defined. Usually, this is done according to the computation model. For example in the data parallel model, the number of data that are processed by a CE constitutes its load. In the task model however, it is the number of tasks that are executed by a CE that defines its load.

9.5.3   Implementation of static algorithms

The network topology has to be defined for each node of the computation. Readers should be reminded that a node can be a machine or a process. The topology is defined by letting each node know its direct neighbors in the computation. For instance, in data parallel applications, the topology is often a result of the data decomposition and mapping phases of the parallelization procedure.

Once the topology is known, the degree of each node is considered to be known and consequently Boillat's diffusion load balancing can be directly implemented. If the Xu's optimized diffusion is to be implemented then optimal diffusion parameters should be computed. This is especially possible when the topology is of the k-ary n-cube class for which Xu et al. found optimal diffusion parameter values.

For the GDE algorithm, however, an edge coloring should be carried out. Network topologies of simple shape like the chain, the ring, the 2D mesh, the torus, the 3D cube and the hypercube are very suitable for ad hoc *a priori* edge coloring. For example, in a chain or ring of size n, two colors are needed. Supposing that each CE has an integer identifier *id*, each processor can assign colors to its incident edges based on its identifier. More precisely, a CE with even *id* has color 0 with its right most edge and color 1 with its left edge. And vice versa for CEs with odd *id*. Similar simple strategies can be defined for the 2D meshes and tori which are colored with four colors. In addition to edge coloring, the optimal exchange parameter value should be computed in the network topologies of the k-ary n-cube class. If the network topology is arbitrary then an edge coloring algorithm should be implemented and run as an initial step before the load balancing algorithm can act. Additionally, no optimal exchange parameter formulae exists for arbitrary networks so the ADE scheme is the simplest approach for setting it. Notice that it may be improved by a test approach and fixed definitely.

Finally, any load balancing algorithm and especially a totally distributed one should be able to detect the balance state. This feature defines its termination detection property. Termination can be implemented in a centralized way or a distributed one. The centralized approach is very simple to implement but carries all the disadvantages of centralized distributed algorithms. Hence a distributed approach is desirable. In GDE (FOS) in particular, a CE usually compares its load with that of its neighbor(s). With GDE, if the difference is less or equal than one then load balance state with that neighbor is detected. With FOS, however, the balance is detected if the difference (the deficiency or overload) is less or equal to the degree of the considered node. Note that this feature is totally distributed.

9.5.4 Implementation of dynamic algorithms

In dynamic networks, some links can come down and recover. So from the point of view of running processes an additional burden should be tackled. The developer should take care of letting processes know what alive links or similarly broken links are. In practice, simple ad hoc dedicated mechanisms or universal libraries that implement sophisticated ones such as failure detectors can be of great help. A dedicated technique can be for example implemented as follows: outgoing messages are sent by a hub function/component that can monitor the delivery of each message to each neighbor. Messages that are not acknowledged or timing out connections can be used as an indicator that the link is broken. This information is then forwarded to the load balancing algorithm that supports link failures. In synchronous implementations, this should take place before each load balancing step.

In the following section, we report an experiment that Vernier achieved out with us during his PhD.

9.6 A practical example: the advection diffusion application

The advection diffusion application solves a kinetic chemicals problem that is described by means of a partial differential equation (PDE). The evolution in space of the concentrations of two chemical species c^1 and c^2 during their reaction is modeled by a PDE which has to be solved in order to trace them. We consider a two dimensional space. A cartesian discretization of this space is made to create a mesh. The PDE is then solved at each point of this mesh in order to find the concentrations of the chemical species. The differential system that permits to find the concentration is given by equation 9.14 where i is used to distinguish the two chemical species

$$\frac{\partial c^i}{\partial t} = K_h \frac{\partial^2 c^i}{\partial x^2} + V \frac{\partial c^i}{\partial x} + \frac{\partial}{\partial z} K_v(z) \frac{\partial c^i}{\partial z} + R^i(c^1, c^2, t), \qquad (9.14)$$

and the terms $R^i(c^1, c^2, t)$ are given by

$$R^1(c^1, c^2, t) = -q_1 c^1 c^3 - q_2 c^1 c^2 + 2q_3(t)c^3 + q_4(t)c^2$$
$$R^2(c^1, c^2, t) = q_1 c^1 c^3 - q_2 c^1 c^2 - q_4(t)c^2$$

When solving the problem the following values are used for the parameters and the constants.

$$K_h = 4.0 \times 10^{-6},$$
$$V = 10^{-3},$$
$$K_v(z) = 10^{-8} exp(z/5),$$
$$q_1 = 1.63 \times 10^{-16},$$
$$q_2 = 4.66 \times 10^{-16},$$
$$c^3 = 3.7 \times 10^{16},$$

and the $q_j(t)$ are defined by:

$$q_j(t) = exp[-a_j/sin(\omega t)] \; for \; sin(\omega t) > 0$$
$$q_j(t) = 0 \qquad\qquad for \; sin(\omega t) \leqslant 0$$

where $j = 3$, 4, $\omega = \pi/43200$, $a_3 = 22.62$, $a_4 = 7.601$. The time interval of the integration is $[0, 7200s]$. The initial Neumann conditions are imposed and are:

$$c^1(x, z, 0) = 10^6 \alpha(x)\beta(z),$$
$$c^2(x, z, 0) = 10^{12} \alpha(x)\beta(z)$$

with

$$\alpha(x) = 1 - (0.1x - 1)^2 + (0.1x - 1)^4/2,$$
$$\beta(z) = 1 - (0.1z - 1)^2 + (0.1z - 4)^4/2$$

The space discretization allows to transform the PDE system into an ODE system of the type

$$\frac{dy(t)}{dt} = f(y(t),\ t) \tag{9.15}$$

that models the equation 9.14. The problem thus turns out to that of solving the ODE given by equation 9.15. By using the implicit Euler method to discretize the equation 9.15, we obtain equation 9.16.

$$\frac{y(t+h) - y(t)}{h} = f(y(t+h), t+h) \tag{9.16}$$

When the function f is not linear, we cannot find the solution of this formula. That is why the fixed point iterative Newton method is used to find the quantities $y(t+h)$. This achieves our brief explanation of the mathematical solution. For a detailed explanation please refer to [Vernier, 2004].

For this problem, the discretization mesh is stored in a vector y. The vector y is defined as :

$$y = (c_{11}^1, c_{11}^2, \ldots, c_{Mx1}^1, c_{Mx1}^2, c_{12}^1, c_{12}^2, \ldots, c_{1Mz}^1, c_{1Mz}^2, \ldots, c_{MxMz}^1, c_{MxMz}^2),$$

where Mx and Mz are respectively the number of elements on the axes x and z of the two dimensions. Thus, a concentration c_{jk}^i, the concentration of the chemical species i at the coordinates jk, is located in the vector y at position $(j - 1 + (k - 1) * Mx) * 2 + i$.

The y_j functions (with $j \in \{1, \ldots, 2 * Mx * Mz\}$) allowing to determine the evolution at each point of each chemical species are consequently also seen as spatial components in the following.

In order to make this model clear, let us take the example of Figure 9.10 where we discretize the considered space in 2 elements on the x axis ($Mx = 2$) and in 2 elements on the z axis ($Mz = 2$). The resulting vector is then constituted of 8 elements (concentrations) noted Ci_jk (where i represents the considered species and jk the coordinates of the element).

Once the vector y is built, the process of solving the ODE system is simply the implementation of the different steps presented above. The algorithm runs in several major time steps, and each one of these steps is staged in three steps: the first one consists in discretizing the ODE with an implicit method of Euler's method, the second one consists in obtaining a linear system with Newton's method and finally the last one aims at solving the resulting linear system. In order to solve the linear system, we use $GMRES$ method (Generalized Minimum RESidual) which is available as part of the $sparseLib++$ [Dongarra et al., 1994] matrix calculus library. The constants of the problem (the time step, the resolution interval, ...) are given above; we also set a threshold ε under which we consider that the linear system has converged. The reader should notice that the solving of our problem is doubly iterative: The first iteration is for the Newton's method and the second for solving the resulting linear system.

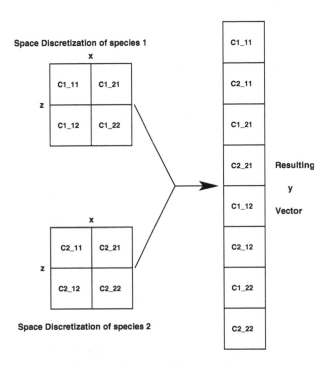

FIGURE 9.10: Space discretization of the mesh.

The distributed solution of this problem needs data partitioning (i.e., decomposition) and mapping. Several methods are possible, the classical ones being a decomposition according to the axes. In our case, we choose to partition according the z axis, because this axis is the simplest with respect to data decomposition and gives us an important advantage for load balancing. Other classical methods consist in decomposing according to the x axis or according to both axis. Figure 9.11 shows how the y vector is partitioned and how its entries are mapped on two 02 CEs. The decomposition that we have chosen implies a communication topology having a chain shape: a processor/machine has always two neighbors, one to the left and the other to the right except for the ones furthest to the left or to the right. So the load balancing is performed on one dimension and is easy to implement. Should the data decomposition have been done according to both axis, the implementation would have been clearly not trivial because the communication topology might have changed.

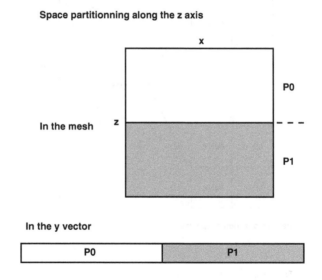

FIGURE 9.11: Partitioning of y and mapping of its elements on different processors.

9.6.1   Load balancing and the application

When integrating a load balancing algorithm to a distributed application, the first thing to do is to define what the load of each CE is. In our case, we chose to define it by the number of elements of the y vector that are computed by a CE. Secondly, the available computation power must be appropriately measured. We choose to compute the average execution time of Newton iterations at the beginning of each major time step.[9] This choice has the advantage of estimating the computation power of the CE available for our application taking naturally into account the nominal computation power of each CE, its memory availability and latency and of background process/external applications. The load migration process will transfer y elements between processors so that the global execution time over the integration interval is minimum. The amount of transfers in turn is determined by the load balancing exchange/diffusion parameter. We recall that the load balancing runs in a chain communication model, i.e., a CE hands off load units for his left and to his right neighbor (by alternating between them with GDE and by considering them both with FOS).

Because we wanted to solve the advection diffusion problem for various discretization steps (by using different values for M_x), we have discarded the methods that need a dynamic looking up step for an optimal relaxation/diffusion parameter (RFOS and SOS). Our choice has been the optimally tuned diffusion algorithm with $\alpha = 0.5$ which is equivalent with ADE on the chain. For this static network context experiment, as we pointed it out, a load balancing frequency should be defined. We chose as we have pointed to compare a load balancing step every major step against a load balancing step every 3 major steps. The distributed application has then been run on a cluster computing environment featuring several workstations with different nominal computing powers (2.4GHz, 2.6Ghz and 1.7GHz) and communicating over a 100 MB/s Ethernet network. The computers were almost dedicated to the computation; only a few not heavy background processes were running with our computation. First, the results of runs on 4 workstations with different powers using different values of M_x featuring the gain of load balancing are depicted in Table 9.3.

Many observations can be made for these results. We will cite some of them here. The first is that the obtained gain is important especially for large sized problems. Secondly, the lazy load balancing frequency makes generally less gain when compared to the rather aggressive frequency. But only generally; when the problem size grows, it tends to have the same gain (cf. the last column). This can be easily explained by the problem size/frequency ratio that enables the LB to be effective. The LB algorithm needs to run a given number of iterations to balance the load. If this number is not reached because the problem size is small (the problem is solved before the LB algorithm can do its job) or the frequency is small than the gain is modest. Otherwise, the LB algorithm proves its usefulness.

---

[9]Consequently we also decide when to take into account the variability of the available computation power.

Mesh size : x x z	100	200	300	400	500	600	700
Without LB	11.4	75.3	266	628	1391	2656	4541
With LB (1)	11	72.7	253	603	1243	2285	3774
Gain with LB (1)(%)	3	3	5	4	10	14	17
With LB 1/3 (2)	10.8	72.8	255	608	1261	2326	3766
Gain with LB(2)(%)	3	3	5	3	9	12	17

Table 9.3: Execution times (s) for the advection-diffusion problem according to the size of the problem with different LB frequencies in a static chain network.

9.6.2   Load balancing in a dynamic network

For a dynamic network context, a lot of changes were introduced to the application, among which the most important is desynchronizing the application (making it asynchronous) if the application has good properties ensuring asynchronous convergence, see [Bertsekas and Tsitsiklis, 1989]. From the programming point of view, this begins by making the communications non-blocking. Indeed, in a dynamic network, if a message is lost then the computations should proceed (a particular CE should not block waiting for a message otherwise it might block indefinitely or at least for a long time). In the particular case of scientific iterative computations, this corresponds to enabling the processors to run different iterations of an iterative algorithm. If a message is received during the computations then its content is integrated on the fly to the current iteration, otherwise older values of data [10] are used instead. In our case, the Newton iterations are run asynchronously but discretized time steps remain synchronous. Notice that a special a priori study must be carried out on the suitability of the asynchronism for the target application – this has been obviously made for the advection-diffusion problem. The suitability is governed by very strict yet realistic conditions that must be met by the application. Finally, the convergence detection of the LB algorithm has been modified to meet the asynchronism of the application. The new algorithm avoids local optima of convergence[11] too.

In a first step, we kept our definitions of the available computation power of a CE but preliminary results showed that the LB algorithm results in poor performance. By analyzing the problem, we pointed out that our definition of the available computation power has to be adapted. We changed it accordingly so that it now reflects the necessary time for a CE to converge definitely in its Newton iterations. Notice also that we use a synchronous LB implementation so there is no need for thresholds to be defined and tuned. The LB iterations

---

[10]For which the not arriving message should have brought new values.

[11]A CE believes the uniform load distribution is reached but it's only temporary.

take place at each major time step.

Results for this asynchronous context are shown in Table 9.4. This shows again the usefulness of load balancing for asynchronous implementations even when no broken edges happen. The gain varies from 10% to more than 25% even if some disturbances can be noticed for meshes greater than 400x400 for reasons that would be too long to explain here.

Problem size : $x \times z$	100	200	300	400	500	600	700
Without LB	22.8	117	396	936	1757	3283	5066
With LB	20	94	307	692	1418	2930	4525
Gain with LB (%)	12	20	22	26	19	11	11

Table 9.4: Execution times (s) of the advection-diffusion problem according to different sizes in an asynchronous implementation.

In the last steps of our work, we introduced some tricks to simulate the temporary link failures. Our asynchronous Advection-diffusion application supports message loss but not definite link failures. But the load balancing algorithm is run synchronously so it does not support message loss but does support link failures. In order to adapt to both we created message delays in the communication layer. This causes some messages to be undelivered which is supported by the asynchronous advection-diffusion and the load balancing algorithm considers links which have delays as broken so it does not send messages to them. In practice, we delayed the messages every two (2) major timesteps. The available computation power is again measured by the time necessary for the Newton iterations to definitely converge.

Problem size : $x \times z$	100	200	300	400	500	600	700
Without LB	45.2	727	1777	2487	4090	7203	9630
With LB	43.9	719	1413	2148	3605	6314	8210
Gain with LB (%)	3	1	20	14	12	12	14

Table 9.5: Execution times of the advection-diffusion problem according to different sizes in an asynchronous implementation on a dynamic network.

We can notice little impact of link failures on the gain obtained by load balancing. If we compare the results for static networks from Table 9.3 to the results of Table 9.5 we can see that the load balancing gain is almost quasi equivalent. But the execution times are very different, this is the raw impact of link failures on the computations.

9.7   Concluding remarks

In this chapter, the latest state of the art load balancing algorithms for static and dynamic networks have been reviewed and discussed. Load balancing has been devised as a technique to turn the heterogeneity of computing elements into an advantage by allowing their efficient use. Originally, LB algorithms run on cluster computing environments which evolved since then to Grid Computing environments. Nowadays, grid computing technologies continue their spread in two directions. In the vertical plane, Grid oriented developing tools and running environments are made accessible and affordable. In the horizontal one, new applications are developed. Current Grid architectures are generally formed by connecting distant clusters by point to point communication links over public networks like the Internet. Computing nodes are generally considered to be highly available but not the links. This context shows the background picture that motivated the design of new load balancing algorithms that support link failures in what is commonly known as dynamic networks.

Nowadays, due to the high cost associated with Grid architecture and the huge popularity of peer-to-peer file sharing systems, a new approach to running computations in a potentially infinite number of machines is receiving more and more attention. Besides, the attraction of peer-to-peer computing raises the same issues of developing tools and running environments. Amongst important missions to carry is new load balancing algorithms that support both link and node failures. An important step into this direction has been already taken with load balancing algorithms that support a variable number of nodes, see [Bahi et al., 2007]. We look forward to implementing and testing these algorithms in computing environments that support node failures like the JACE (Java Asynchronous Computing Environment) environment. JACE is a java library that permits easy programming of grid aware scientific distributed software. It supports (non-) blocking communications and sophisticated distributed convergence detection, i.e., important things to make a distributed application support dynamic networks.

9.8 References

[Aiello et al., 1993] Aiello, W., Awerbuch, B., Zkfaggs, B., and Rao, S. (1993). Approximate load balancing on dynamic and asynchronous networks. In *Proceedings of the 25th Annual ACM Symposium on Theory of Computing*, pages 632–641.

[Bahi et al., 2006a] Bahi, J., Couturier, R., and Sider, A. (2006a). Design and analysis of the M2LL policy distributed algorithm for load balancing in dynamic networks. In *Proceedings of the 2006 International Parallel and Distributed Processing Symposium (IPDPS 2006)*, volume 4331 of *Lecture Notes in Computer Sciences*, pages 195–204, Heidelberg. Springer-Verlag.

[Bahi et al., 2003a] Bahi, J., Couturier, R., and Vernier, F. (2003a). Accelerated diffusion algorithms on general dynamic networks. In *Proceedings of 5th International Conference (PPAM)*, volume 3019 of *Lecture Notes in Computer Sciences*, pages 77–82, Czestochowa, Poland. Springer-Verlag.

[Bahi et al., 2003b] Bahi, J., Couturier, R., and Vernier, F. (2003b). Broken edges and dimension exchange algorithm on hypercube topology. In *Proceedings of the 11th Euromicro Conference on Parallel, Distributed and Network-Based Processing (Euro-PDP'03)*.

[Bahi et al., 2005] Bahi, J., Couturier, R., and Vernier, F. (2005). Synchronous distributed load balancing on dynamic networks. *Journal of Parallel and Distributed Computing*, 65:1397–1405.

[Bahi et al., 2007] Bahi, J., Couturier, R., and Vernier, F. (2007). Synchronous distributed load balancing on totally dynamic networks. In *Proceedings of the 2007 International Parallel and Distributed Processing Symposium (IPDPS 2007)*, pages 1–8. IEEE International.

[Bahi et al., 2006b] Bahi, J., Couturier, R., and Vuillemin, P. (2006b). Solving nonlinear wave equations in the grid computing environment: an experimental study. *Journal of Computational Acoustics*, 14(1):113–130.

[Bahi and Gaber, 2001] Bahi, J. and Gaber, J. (2001). Load balancing on networks with dynamically changing topology. In *Proceedings of the 7th International Europar Conference on Parallel Processing*, pages 175–182, Manchester. Lecture Notes in Computer Sciences.

[Bertsekas and Tsitsiklis, 1989] Bertsekas, D. and Tsitsiklis, J. (1989). *Parallel and distributed computation: numerical methods*. PrenticeHall, Englewood Cliffs, NJ, USA.

[Boillat, 1990] Boillat, J. (1990). Load balancing and Poisson equation in a graph. *Concurrency: Practice and Experience*, 2(4):289–313.

[Cortes et al., 1999] Cortes, A., Ripoll, A., Senar, M., and Luque, E. (1999). Performance comparison of dynamic load-balancing strategies for distributed systems. In *Proceedings of the 32th Hawaii International Conference on System Sciences*, volume 8, pages 8041–8051. IEEE Computer Society.

[Cybenko, 1989] Cybenko, G. (1989). Dynamic load balancing for distributed memory multiprocessors. *Journal of Parallel and Distributed Computing*, 7(2):279–301.

[Diekmann et al., 1999] Diekmann, R., Frommer, A., and Monien, B. (1999). Efficient schemes for nearest neighbor load balancing. *Parallel Computing*, 25(7):789–812.

[Dongarra et al., 1994] Dongarra, J., Lumsdaine, A., Pozo, R., and Remington, K. (1994). A sparse matrix library in C++ for high performance architectures. In *Proceedings of the 2nd Object Oriented Numerics Conference*, pages 214–218.

[Elsasser et al., 2002] Elsasser, R., Monien, B., and Preis, R. (2002). Diffusion schemes for load balancing on heterogeneous networks. *Theory of Computing Systems*, 35:305–320.

[Elsasser et al., 2004] Elsasser, R., Monien, B., and Schamberger, S. (2004). Load balancing in dynamic networks. In *Proceedings of the 7th International Parallel and Distributed Processing Symposium (I-SPAN'04)*, pages 193–200. IEEE Computer Society.

[Fiorini and Wilson, 1978] Fiorini, S. and Wilson, R. (1978). Edge-coloring of graphs. In Beineke, L. and Wilson, R., editors, *Selected topics in graph theory*, New York, USA. Academic Press.

[Geist et al., 1994] Geist, A., Beguelin, A., Dongarra, J., Jiang, W., Manchek, R., and Sunderam, V. (1994). *PVM: a users' guide and tutorial for networked parallel computing*. MIT Press.

[Ghosh et al., 1996] Ghosh, B., Muthukrishnan, S., and Schultz, M. (1996). First and second order diffusive methods for rapid, coarse, distributed load balancing. In *Proceedings of the 8th Annual ACM Symposium on Parallel Algorithms and Architectures*, pages 72–81, Padua, Italy. Sigact/Sigarch.

[Gropp et al., 1994] Gropp, W., Lusk, E., and Skjellum, A. (1994). *Using MPI: portable parallel programming with the message passing interface*. MIT Press.

[Henrich, 2004] Henrich, D. (2004). Local load balancing according to a simple liquid model. Technical report, Institute for Real-Time Computer Systems and Robotics, University of Karlsruhe, D-76128 Karlsruhe, Germany.

[Hosseini et al., 1990] Hosseini, S., Litow, B., Malkawi, M., McPherson, J., and Vairavan, K. (1990). Analysis of a graph coloring based distributed load balancing algorithm. *Journal of Parallel and Distributed Computing*, 10(2):160–166.

[Kumar et al., 1991] Kumar, V., Ananth, G., and Rao, V. (1991). Scalable load balancing techniques for parallel computers. Technical report, Department of Computer Science, University of Minnesota, USA.

[Lin and Keller, 1987] Lin, F. and Keller, R. (1987). The gradient model load balancing method. *IEEE Transactions on Software Engineering*, 13(1):32–38.

[Rotaru and Nageli, 2004] Rotaru, T. and Nageli, H. (2004). Dynamic load balancing by diffusion in heterogeneous systems. *Journal of Parallel and Distributed Computing*, 64:481–497.

[Sanders, 1994] Sanders, P. (1994). Analysis of random polling dynamic load balancing. Technical report, Lehrstuhl Informatik Ingenieure und Naturwissenschaftler. University of Karlsruhe, D–76128 Karlsruhe, Germany.

[Sider and Couturier, 2009] Sider, A. and Couturier, R. (2009). Fast load balancing with the most to least loaded policy in dynamic networks. *Journal of SuperComputing*. (in press).

[Varga, 1962] Varga, R. (1962). *Matrix iterative analysis*. Automatic Computations. PrenticeHall, Englewood Cliffs, NJ, USA.

[Vernier, 2004] Vernier, F. (2004). *Algorithmique itérative pour l'équilibrage de charge dans les réseaux dynamiques*. PhD thesis, Université de Franche-Comté, France.

[Willebeek-LeMair and Reeves, 1990] Willebeek-LeMair, M. and Reeves, A. (1990). Local versus global strategies for dynamic load balancing. In *Proceedings of the International Conference on Parallel Processing*, pages 569–570.

[Willebeek-LeMair and Reeves, 1993] Willebeek-LeMair, M. and Reeves, A. (1993). Strategies for dynamic load balancing on highly parallel computers. *IEEE Transactions on Parallel and Distributed Systems*, 4–9:979–993.

[Xu and Lau, 1992] Xu, C. and Lau, F. (1992). Analysis of the generalized dimension exchange method for dynamic load balancing. *Journal of Parallel and Distributed Computing*, 16(4):385–393.

[Xu and Lau, 1994] Xu, C. and Lau, F. (1994). Optimal parameters for load balancing with the diffusion method in mesh networks. *Parallel Processing Letters*, 4(2):139–147.

[Xu and Lau, 1996] Xu, C. and Lau, F. (1996). *Load balancing in parallel computers: theory and practice.* Kluwer Academic Publishers, Boston, MA, USA.

[Xu et al., 1995] Xu, C., Monien, B., Lüling, R., and Lau, F. (1995). Nearest neighbor algorithms for load balancing in parallel computers. *Concurrency: Practice and Experience*, 7:707–736.

Appendix A

Implementation of the replication strategies in OptorSim

Thi-Mai-Huong Nguyen

Applied Mathematics and Systems Laboratory, Ecole Centrale Paris, Grande Voie des Vignes, 92295 Châtenay-Malabry, France

Frédéric Magoulès

Applied Mathematics and Systems Laboratory, Ecole Centrale Paris, Grande Voie des Vignes, 92295 Châtenay-Malabry, France

A.1 Introduction

Three replication strategies proposed in the Chapter 3, i.e., MaxDAR-Pb, MaxDAR-Pz and MaxDAR-C, based on the MaxDAR algorithm were implemented using the grid simulator OptorSim v2.0.1. In this appendix, we present the implementation of the replication strategies in OptorSim and give a brief example on how the simulation is executed to evaluate them.

The grid simulator OptorSim is written in the Java programming language and originally developed as part of the European DataGrid (EDG) project. The OptorSim provides APIs for simulation of any grid topology and job execution by means of a few configuration files. Several job scheduling and file replication algorithms are implemented in OptorSim as replica optimizers and more can easily be added. We have implemented three MaxDAR optimizers as three new replica optimizers in OptorSim. The detailed architecture of OptorSim was introduced in Section 3.3.

A.2   Download

Please see the section "Downloads" on `http://sourceforge.net/projects/optorsim` for source download or contact directly the authors.

A.3   Implementation

A.3.1   OptorSim implementation

There are two different types of optimization which may be investigated using OptorSim: the scheduling algorithms used by the resource broker to allocate jobs, and the replication algorithms used by the replica manager at each site to decide when to replicate a file, which file to replicate and which to delete. The overall aim is to reduce the time it takes jobs to run, and also to make the best use of grid resources. In the short term, an individual user wants their job to finish as quickly as possible, but in the long term the goal is to have the data distributed in such a way as to improve job times for all users, thus giving the greatest throughput of jobs. The scheduling algorithm and replication strategies currently implemented are listed in Table A.1.

Scheduling	Replication
Random - schedule to random site	*No replication*
Access Cost - site where time to access all files required by job is shortest	*Least Recently Used (LRU)* - always replicate, delete least recently used file
Queue Size - site where job queue is shortest	*Least Frequently Used (LFU)* - always replicate, delete least frequently used file
Queue Access Cost - site where access cost for all jobs in queue is shortest	*Economic model (Binomial)* - replicate if economically advantageous, using binomial prediction function for file values
	Economic model (Zipf) - replicate if economically advantageous, using Zipf-based prediction function

Table A.1: Scheduling algorithm and replication strategies implemented in OptorSim.

Each scheduling and replication algorithm is implemented as a separate resource broker or replica optimizer class respectively and the appropriate class is instantiated at runtime, making the code highly extensible. In OptorSim, each computing element is represented by a thread, with another thread acting as the resource broker. The resource broker sends jobs to the computing elements according to the specified scheduling algorithm and the computing elements process the jobs by accessing the required files, running one job at a time until they have finished all their jobs.

A.3.2 MaxDAR implementation

Figure A.1 shows the class diagram of implemented replica optimizers including MaxDAR replication strategies which are proposed in this appendix.

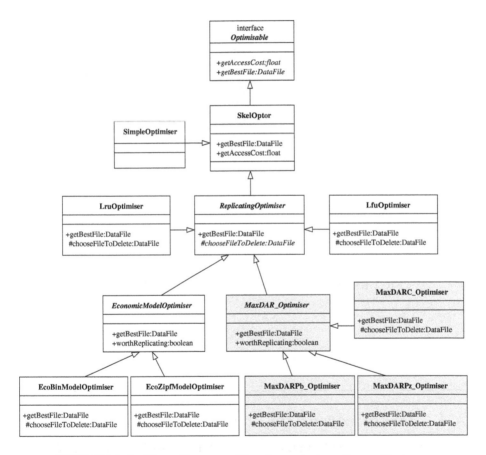

FIGURE A.1: Class diagram of implemented replica optimizers.

Figure A.2 illustrates the sequence diagram of a computing element's file request. When a file is needed, the computing element calls the *getBestFile()* method of the replica optimizer being used. The replication algorithm is then used to search for the "best" replica to use, and the file is either replicated to the local site or read remotely. In case there is not enough free space, the replica optimizer determines whether the data should be replicated to local storage and which replicas should be removed by calling the *choose-FilesToDelete()* method. This method invokes the *filesToDelete()* method of MaxDAR optimizer which in turn evaluates the benefits and storage cost of the chosen files to be deleted by calling the *getProfit()* and *getStorageCost()* methods. These two methods use the *evaluateRank()* method which returns the rank of the file according to its popularity or correlation with other files. The replica optimizer removes the chosen files for the new replica if the replication benefit is greater than the replacement loss.

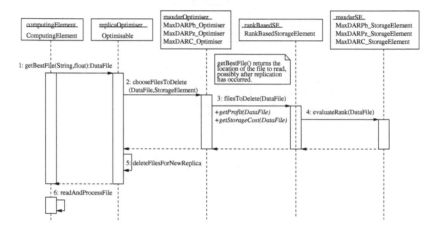

FIGURE A.2: Sequence diagram of a CE's file request.

A.4 How to execute the simulation

OptorSim can be run from the command line to provide a number of statistics:

- Total and individual job times

- CE usage

- Number of replications, local and remote file accesses

- SE usage

The appropriate statistics are output on the level of the grid, individual sites and site components. In order to run the simulation, the class `org.edg.data.replication.optorsim.OptorSimMain` needs to be launched using the configuration file included in the `conf` directory. The output of the simulation performed in this chapter is placed in the `samples/data` directory.

Appendix B

Implementation of the simulator for the distributed scheduling model

Lei Yu

Applied Mathematics and Systems Laboratory, Ecole Centrale Paris, Grande Voie des Vignes, 92295 Châtenay-Malabry, France

Frédéric Magoulès

Applied Mathematics and Systems Laboratory, Ecole Centrale Paris, Grande Voie des Vignes, 92295 Châtenay-Malabry, France

B.1   Introduction

The simulator is programmed in C and it is composed of three parts of functional files. The first is `masterslaver1.c` which creates the simulation environment and all the needed hosts. The `sched_struct.c` and `gestemessage.c` provides the basic functions of the simulation. Last, the `ma.c`, `la.c`, `sla.c`, `sed.c`, `calculer.c` and `client.c` simulate the functionality and algorithm of each type of server in the distributed scheduling system.

B.2   Download

Please see the section "Downloads" on `http://sourceforge.net/projects/simulator` for source download or contact directly the authors.

B.3   Implementation

B.3.1   Data structures

In order to facilitate the programming, several data structures are defined:

- `channel_t`: The names of the channels used in the simulation.

- `s_PSL(*)`: The structure type of LSP that can be used to construct the tree structure of scheduling system. When a child of the failed scheduler detects the failure of its father, the child creates the tree structure of system, and uses this tree structure to find out its new father.

- `s_PSL_seri(*)`: The serialized LSP. It is transferred as a message between schedulers.

- `s_PTC_seri(*)`: The serialized CTP which is created by *ma* using the received `s_PSL_seri_t` from its children, and is transferred regularly to all its children.

- `s_Job_status(*)`: The structure to maintain job state information.

- `s_message(*)`: It is the structure that is used to create messages needed to be transferred.

B.3.2   Functions

B.3.2.1   Messages operations

- `s_message_tcreate_message(*,*,*,*,*,*,*,*)`: Create the `s_message` structure and return a `s_message_t`.

- `s_message_tcopy_message(*)`: Create a copy of message.

- `voiddestroy_message(*)`: Destroy message structure and free the allocated memory to the operating system.

- `intprocess_send_task(*,*)`: Define the sending message operations and is used by `MSG_process_create` to create a process to send a message

- `s_message_tget_message(*,*)`: Get message from a channel. The massage can be got in the manner synchronous or asynchronous.

B.3.2.2 Simulator defined structures operations

- `s_PSL_seri_tPSL_seri_data_create(*)`: Create a `s_PSL_seri` structure.

- `s_PSL_seri_tPSL_seri_data_copy(*)`: Create a copy of `s_PSL_seri` structure.

- `voidPSL_seri_data_destroy(*)`: Destroy a `s_PSL_seri` structure.

- `s_PTC_seri_tPTC_seri_data_create(void)`: Create a `s_PTC_seri` structure.

- `s_PTC_seri_tPTC_seri_data_copy(*)`: Create a copy of `s_PTC_seri` structure.

- `voidPTC_seri_data_destroy(*)`: Destroy a `s_PTC_seri` structure.

- `s_PSL_tPSL_data_create(*)`: Create a `s_PSL` structure.

- `s_Job_status_tJob_data_create(*,*,*)`: Create a `s_Job_status` structure.

- `s_Job_status_tJob_data_copy(*)`: Create a copy of `s_Job_status` structure.

- `s_PTC_seri_tPTCseri_create_from_topologie_locale(*)`: Create a `s_PTC_seri` structure and initiate the structure according to all its children's `s_PSL_seri`.

- `s_PSL_seri_tPSLseri_create_from_topologie_locale(*,*)`: Create a `s_PSL_seri` structure and initiate the structure according to all its children's `s_PSL_seri`.

B.3.2.3 Tree structure operations

- `s_PSL_ttopologie_create_from_PTCseri(*)`: Create the tree structure of system and return a `s_PSL_t` as the root of the tree.

- `s_PSL_tfind_me(*,*)`: Find the host itself in the tree structure.

- `m_host_tget_host_from_ID(*,)`: Find the host in the tree structure according its ID.

- `voiddestroy_tree(*)`: Destroy the tree structure and free the allocated memory to the operating system.

B.3.2.4 Applications deployment operations

- `int*create_application_list_from_PSLseri(*,*)`: Create the deployed applications list from the `s_PSL` structure.

- `int*get_application_array(*,*,*)`: Get the deployed applications list in a host.

- `intis_application_found(*,*,*)`: Find whether an application is in the applications list.

B.3.2.5 Miscellaneous operations

- `voidadd_son(*,*)`: Insert a child's `s_PSL` into its children's list.

- `intdel_son(*,*)`: Delete a `s_PSL` from its children's list.

- `voidjob_info_print(*)`: Print job information according to the job's complete sequence.

- `voidjob_average_print(*)`: Print average information of jobs according to the job's type.

B.4 How to execute the simulation

The simulator is developed in language C based on the top of the toolkit SimGrid. The installation and execution of this simulator must be effectuated after the installation and configuration of SimGrid. This simulator implemented the simulation of a distributed dynamic scheduling algorithm with fault tolerance (DDFT); the users can modify the source to achieve their own scheduling algorithm.

In the directory source, a make file is created. The users must first set the INCLUDES and DEFS variables to the path of installed SimGrid toolkit in the makefile. Then the users can use the command make to compile the source and create the execution file, lance, of the simulator. After having created the file lance, the command `./lanceplatform.xmldeployment.xml` is used to start the simulation. In the directory sample, a `platform.xml` file and a `deployment.xml` file have been saved as the samples of this simulation. At the end of simulation, the experimental results will be shown in a table format.

Glossary

A

Abstract job object (AJO)   is a Java object that allows users to define jobs independent from the system. The jobs created by the clients are encapsulated as AJO.

Aggregate directory   is a collective repository for the resources present in the grid. GRIS and GIIS are examples of aggregate directories in MDS.

Aggregator services   are the services built on top of the aggregator framework that use aggregator sources to collect data. They can be queried to find information about resources in the grid using XPath queries.

Aggregator sources   are Java classes that are part of the aggregator framework of MDS. They implement an interface to collect XML formatted data from registered information providers (IPs).

AHEFT   is a heterogeneous earliest finish time-based adaptive rescheduling strategy for grid workflows. It reschedules the jobs in the workflow by monitoring the performance of the jobs in the workflow and by discovering newly available resources in the grid.

Authentication   is the process by which an entity establishes its identity to the other entities in the network.

Authorization   is the verification of the privileges assigned to an entity to access the resources and services provided by other entities in the grid.

B

Backfill algorithm   is a scheduling algorithm that tries to find a job that can be started with the current available resources if the job at the head of the queue cannot be started due to lack of resource availability. This should not delay the scheduled start of the job at the head of the queue.

Berkeley database information index (BDII) is an LDAP server that gathers information from individual GIIS. It contains a grid-level view of all the resources available in the grid.

C

Certificate revocation list (CRL) is a list of the serial numbers of the X.509 digital certificates that cannot be trusted because their validity has ended or because of some fraud.

Certifying authority (CA) is a trusted third party in the PKI that issues digital certificates to individuals and organizations.

Chain of trust is the process of trust establishment between an entity and a CA. This is done by verifying the correctness of the public key of the CA by tracing it upwards to another CA in the PKI hierarchy trusted by the entity.

Communication overhead is the additional processing time spent by the system for control checking and error checking. For parallel computation, data exchange between the independent nodes constitutes the communication overhead.

Computing element (CE) is a grid resource that carries out the execution of a job.

Condor pool is a collection of agents, resources, and matchmakers.

Confidentiality refers to the hiding of sensitive information from the entities that do not have the rights to access them. It can be done either at the message level or at the transport level.

Credential delegation is the process of delegating one's complete or partial privileges to another entity in the grid. This allows the entity to access the resources on the behalf of the entity delegating the credential. Proxy certificates are used for credential delegation.

D

Data replication service (DRS) is a service to provide a pull-based replication capability for grid files. It is a high-level data management service built on top of two GT data management components:

the Reliable File Transfer (RFT) Service and the Replica Location Service (RLS).

Directed acyclic graph (DAG) is a graphical representation of dependencies among tasks in a grid workflow. The nodes of a DAG represent the tasks and the directed edges represent the data dependencies.

Directed acyclic graph manager (DAGMan) is a meta-scheduler for the execution of programs (computations) in Condor.

Directory information tree (DIT) is a tree-based structure to organize names of entities in MDS in a hierarchical fashion.

Distributed fault-tolerant scheduling (DFTS) is a fault-tolerance mechanism for grids based on job replication.

E

Expected completion time is the wall-clock time at which a machine completes the execution of a task.

Expected execution time (EET) is the estimated time for the execution of a task on a machine when the machine has no job to execute.

Expected time to compute (ETC) matrix contains the expected execution time of tasks on all the machines in the grid. It is used by the scheduler to make mapping decisions.

Extensible markup language (XML) is a markup language whose purpose is to facilitate sharing of data across different interfaces using a common format.

External data representation (XDR) is a standard for the description and encoding of data. It is used to transfer data between different computer architectures.

F

Fast greedy is a mapping heuristic that assigns tasks to machines in an arbitrary order having the minimum completion time for that task. It is also known as MCT heuristic.

File transfer service (FTS) is a grid component that facilitates the transfer of data between different storage elements in the grid.

G

Genetic algorithm heuristic is a mapping heuristic that uses a genetic algorithm to find the best schedule for a metatask.

Globus access to secondary storage (GASS) allows access to data stored in a remote filesystem. Its client libraries allow applications to access remote files and its server component allows any computer to act as a limited file server.

Globus gridmap file contains a list of global names of users who are authorized to access a service on that node. All authorized users are mapped to a local user.

Globus resource allocation manager (GRAM) processes the requests for resources for remote application execution, allocates the required resources and manages the active jobs. It also returns updated information regarding the capabilities and availability of the computing resources to the Metacomputing Directory Service (MDS).

Grid fabric is a layer containing the grid resources such as computional power, data storage, sensors and network resources.

Grid index information service (GIIS) is a higher-level aggregate directory that collects information about grid resources from GRIS and lower-level GIIS. It contains the grid-level view of the resources.

Grid information protocol (GRIP) is a protocol for the discovery of new grid resources and enquiry of known grid resources. MDS uses LDAP for GRIP.

Grid monitoring architecture (GMA) is a consumer-producer based architecture for information and monitoring services in grids. It contains a registry for storing information about consumers and producers. It forms the basis of R-GMA.

Grid portal is an interface to a grid system. Users interact with the portal using an intuitive interface through which they can view files, submit and monitor jobs and view accounting information.

Grid registration protocol (GRRP) is used by various components of MDS such as IPs and GRIS to inform other components about its existence.

Grid resource information service (GRIS) is a lower-level aggregate directory that collects information about grid resources from the information providers.

Grid resources are the components of a grid that are used in the processing of a job, e.g., computing element, storage element, etc.

Grid security infrastructure file transfer protocol (GSIFTP) is a file transfer protocol built on top of the grid security infrastructure. It allows secure data transfer between grid components.

Grid security infrastructure (GSI) is a part of the Globus Toolkit and defines the necessary standards for the implementation of security in grids. It consists of X.509 digital certificates, SAML and Globus gridmap file, X.509 proxy certificates and message protection using TLS or WS-security and WS-secure conversation.

Grid service is a stateful web service to make it suitable for grid applications.

Grid service handle (GSH) is used to distinguish different grid service instances of the same service created by the factory.

Grid service reference (GSR) contains grid service instance-specific information such as protocol binding, method definition and network address.

Grid workflow is the automation of a collection of services (or tasks) by coordination through control and data dependencies to generate a new service.

Grid workflow management system (GWFMS) is software used for modeling tasks and their dependencies in a workflow and managing the execution of workflow and their interaction with other grid resources.

GridFtp is a protocol defined by global grid forum-based on FTP. It provides secure, robust, fast and efficient transfer of bulk data in the grid environment. The Globus Toolkit provides the most commonly used implementation of the protocol.

GridRPC is a programming model based on client-server remote procedure call (RPC).

H

Heterogeneous computing (HC) refers to a system in which diverse resources are combined together to increase the combined performance and cost-effectiveness of these resources. Grid is an example of an HC system.

High-level Petri-net (HLPN) provides an extension to the classical Petri-nets by adding support to model data, time and hierarchy. It allows computation of output tokens of a transition based on multiple input tokens contrary to classical Petri-nets, which allow only one type of token.

High performance storage system (HPSS) is software that manages peta-bytes of data on disks and robotic tape libraries.

I

Information provider (IP) is a service that interfaces a data collection service and provides information about the available resources to the aggregate directories.

Information service is one of the main components of the grid. It provides static and dynamic information about the grid resources.

Information supermarket is a component in the workload manager that stores information about all active resources in the grid, which are used by the matchmaker for the decision-making process.

Interlogger helps to propagate the logging or book-keeping information from a grid component to the central book-keeping server.

J

Job is a computational task that is executed on the grid. The information pertaining to a job is specified by the user using a job description language.

Job description language (JDL) is a computer language used for describing a job based on information specified by the user.

Job handler is a component in the workload manager that carries out the packaging, submission, cancelation and monitoring of a job.

Job replication is a fault-tolerance strategy in which more than one copy of the same job are assigned to a different set of resources. The different instances may either run the same copy of the job or a copy using a different algorithm for the same job.

Job submission description language (JSDL) is a language to describe the requirements of a job for submission to grid resources. The job requirement is specified using XML.

L

LDAP is a protocol for querying and modifying directory services running over TCP/IP. LDAP support is implemented in web browsers and e-mail programs, which can query an LDAP-compliant directory. LDAP is a sibling protocol to HTTP and FTP and uses the `ldap://` prefix in its URL.

Level-based scheduling algorithm is a scheduling algorithm, which partitions the DAG into levels of independent nodes and then uses heuristics like min-min, max-min and sufferage to map these nodes to processors.

List scheduling algorithm is a scheduling algorithm, which assigns priorities to nodes in a DAG and considers the nodes with higher priority for scheduling before the lower priority nodes.

Local files catalog (LFC) is a grid component that stores the mapping between different identifiers of a file or a resource.

LSF is software for managing and accelerating batch workload processing for computationally intensive and data-intensive applications.

M

Machine availability time (MAT) is the earliest time when a machine has completed the execution of all the previously assigned tasks and is ready to serve the next request.

Makespan is defined as the maximum time taken for the completion of all the tasks in the metatask or for the execution of the complete grid workflow.

Mapper is the component of a grid scheduler, which runs the mapping algorithm.

Mapping is the overall process of matching and scheduling.

Masterworker is a model for the execution of parallel applications in which a node (controlling master) sends pieces of work to other nodes (workers). The worker node performs the computation and sends the result back to the master node. A piece of work is assigned to the first worker node that becomes available next.

Matching is the process of identifying suitable machines for a task.

Matchmaker is a component that performs the matching of a job to grid resources based on the user information on the job and the available information on grid resources.

Max-min is a mapping heuristic that finds the minimum expected completion time for each task in a metatask and then assigns the task having the maximum expected completion time to the corresponding machine.

MCT see fast greedy.

Message passing interface (MPI) is a library of subroutines for handling communication and synchronization of programs running on parallel platforms.

MET see user-directed assignment.

Metacomputing uses many networked computers together as a single computational unit to provide massive processing power.

Metatask is a collection of independent tasks mapped to a collection of resources during a mapping event.

Middleware is a collection of software and packages used for the implementation of a grid.

Min-min is a mapping heuristic that finds the minimum expected completion time for each task in a metatask and then assigns the task having the least expected completion time to the corresponding machine.

Mixed-machine system is a class of HC system, which consists of heterogeneous machines connected by high-speed networks.

Monitoring and discovery services (MDS) is a component of the Globus Toolkit, which provides resource monitoring and discovery services within the grid environment.

MyProxy is an open-source software used for managing user X.509 certificates. It can be used to store and retrieve user credentials over the network in a secure way.

N

Namespace is a naming context in which each name should be unique.

Network job supervisor (NJS)   is one of the Unicore components. It translates the jobs represented as AJO into target system-specific batch jobs. It also passes sub-AJOs to peer systems, synchronizes the execution of dependent jobs and manages the data transfer between different systems.

Node   is a portion of the grid where a job can be executed independently on the grid. The parallel structure of the grid comes from running in parallel jobs on different nodes simultaneously.

O

Open grid services architecture (OGSA)   defines a web services-based framework for the implementation of grid.

Open grid services infrastructure (OGSI)   is a formal and technical specification of the implementation of grid services as defined by the OGSA framework.

Opportunistic load balancing (OLB)   is a mapping heuristic that assigns tasks to the next available machine.

P

Parallel virtual machine (PVM)   is a software package that permits a heterogeneous collection of Unix and/or Windows computers hooked together by a network to be used as a single large parallel computer.

Parameter sweep application   is an application that executes multiple instances of a program using different sets of parameters and then collects the results from all the instances. Such applications frequently occur in scientific and engineering problems.

Petri-net   is a modeling language that graphically represents the state of workflow in grids or distributed systems using the concept of tokens. It consists of places, transitions and directed arcs connecting the places to the transitions.

Pluggable authentication module (PAM)   is a mechanism to integrate low-level authentication schemes with a high-level API so that the application may be written independent of the underlying authentication scheme. For example a MyProxy server can be configured to use an external authentication like an LDAP server.

Portable batch system (PBS)   is a batch job and computer system resource management package. It accepts batch jobs (shell scripts with control attributes) and stores the job until it is run. It runs the job and delivers the output back to the user.

Portlet   is a pluggable user interface component that is managed and displayed in a web portal.

Principal   is the entity whose identity is being verified.

Proxy certificate   is a part of the GSI, which is used by an entity to delegate its complete or partial privileges to another entity. It is also used for single sign-on. It has the same format as an X.509 digital certificate.

Public key infrastructure (PKI)   is a method of secure communication between two entities in the Internet using the public/private key pair. It consists of a trusted third party called the Certifying Authority (CA).

R

Relational grid monitoring architecture (R-GMA)   is an information and monitoring system for grids developed by the European Data-Grid project. It is based on GMA and derives its flexibility from the relational model.

Reliable file transfer (RFT)   is a web service that provides interface for controlling and monitoring third party file transfers using GridFTP. The client controlling the transfer is hosted inside a grid service so that it can be managed using the soft state model and queried using the ServiceData interface available to all grid services.

Remote procedure call (RPC)   is a protocol that allows a program running on one host to invoke a procedure on a different host in the network.

Replica location service (RLS)   is a service that allows the registration and location of replicas in Globus. It maps the logical file name to a physical file name.

Rescheduling   is the process of assigning a job to a new machine, either to improve its performance or for the purpose of fault tolerance.

S

Scheduling   is the process of ordering the execution of a collection of tasks on a pool of resources.

Secure socket layer (SSL)   is a protocol for secure communication over the Internet. SSL uses a public/private key pair for the encryption and decryption of the data. The public key is known to everyone and the private or secret key is known only to the recipient of the encrypted message.

Security assertion markup language (SAML)   is an XML-based standard protocol that supports the exchange of identity information under different environments. Identity information is exchanged as assertions between the provider and consumer of assertions.

Service level agreement (SLA)   defines the minimum quality of sevice, availability and other service-related attributes expected by the user from the service provider and the charges levied on them.

Simple object access protocol (SOAP)   is an XML-based communication protocol, which can be used by two parties communicating over the Internet.

Single program multiple data (SPMD)   is a style of parallel programming where all the processors use the same program but process different data.

Single sign-on   is the process of authenticating once to obtain proxy credentials, which can be used to access grid resources without needing further authentication for a certain period.

Storage element (SE)   is a grid resource that stores the information required or generated by the computing element.

Strike price   is the predetermined price of the underlying stock at expiry date. This price is predetermined at time 0, which is the time when the option is bought.

Sufferage heuristic   is a mapping heuristic that maps tasks in the decreasing order of sufferage value.

Sufferage value   is the difference between best and second-best minimum completion time for a task.

Switching algorithm   is a mapping heuristic that tries to strike a balance between MCT and MET heuristics by switching between the heuristics based on the load of the system.

T

Task farming is a type of parallel application, where many independent jobs are executed on machines around the world. Only a small amount of data needs to be retrieved from each of these jobs.

Task-level fault tolerance achieves fault tolerance by either rescheduling the job or by using a job replication strategy without affecting the workflow.

Task queue is a component of the workload manager that holds various jobs for the eventual allocation by the matchmaker. Authentication of the user information in the job is also done in the task queue.

Testbed is an experimental platform including dedicated hardware, software resources and scientific instruments. It is used to test and analyze the tools and products. It usually supports real-time deployment and interaction.

Ticket granting ticket (TGT) is issued by the AS to a client. The TGS verifies the validity of the TGT before issuing actual communication ticket to the client.

Trust is a relationship between two entities that forms the basis for the subsequent authentication and authorization between the two entities.

Trusted third party (TTP) is an entity that provides for the authentication of two parties, both of which trust the third party. CA is an example of a TTP.

U

Universal description, discovery and integration (UDDI) is an XML-based registry used for finding a web service on the Internet.

User-directed assignment is a mapping heuristic that assigns tasks in an arbitrary order to the machine having the minimum execution time for that task.

V

Verifier is the entity that verifies the identity of the principal.

Virtual organization (VO) is a dynamic collection of multiple organizations that provides coordinated resource sharing. A grid usually consists of multiple virtual organizations.

W

Web service definition language (WSDL) is an XML document used to describe a web service interface.

Web service (WS) is a software system designed to support interoperable machine-to-machine interaction over a network.

Workflow-level fault tolerance allows changes to the workflow structure to achieve fault tolerance. These include user-defined exceptions and task crash failures that cannot be handled by the task-level failure handling techniques.

Workload manager (WM) is an interface in gLite that deals with the allocation, collection and cancellation of a job. It also provides information about the job status and the grid resources.

WS-federation is a specification for standardizing how organizations share user identities in a heterogeneous authentication and authorization system.

WS-policy is a specification for the service requestor and service provider to enumerate their capabilities, needs and preferences in the form of policies.

WS-privacy is a proposed web service specification. It will use a combination of WS-security, WS-policy and WS-trust for communicating privacy policies among organizations.

WS-resource framework (WSRF) is a generic and open framework for modeling and accessing stateful resources using web services. It contains a set of six web services specifications that define what is termed as the WS-resource approach to model and manage stateful resources in a web service context.

WS-secure conversation is a web service extension built on top of WS-security and WS-trust. It provides a security context for the protection of more than one related message.

WS-security is the standard to provide security features such as integrity, privacy, confidentiality and single message authentication to SOAP messages.

WS-trust is an extension to the WS-security specification. It defines additional constructs and primitives for the request and issue of security tokens. It also provides ways to establish trust relationships with parties in different trust domains.

X

X.509 digital certificate is a standard for digital certificates described by the RFC 2459. It consists of the public key of the certificate owner and is signed by the certifying authority.

XML digital signature is a way of digitally signing the SOAP messages to ensure their integrity.

XML encryption is a standard that provides end-to-end security for applications requiring secure XML data exchange. A SOAP message body is encrypted using block encryption algorithms like AES-256.

XSufferage heuristic is a modification of sufferage heuristics that takes into account the location of data, while making the scheduling decision. Instead of grid-level MCT, XSufferage heuristics use cluster-level MCT to find the sufferage value.

Author Index